AF251781

How to Become an IT Architect

How to Become an IT Architect

Cristian Bojinca

ARTECH HOUSE

BOSTON | LONDON
artechhouse.com

Library of Congress Cataloging-in-Publication Data
A catalog record for this book is available from the U.S. Library of Congress.

British Library Cataloguing in Publication Data
A catalogue record for this book is available from the British Library.

Cover design by John Gomes

ISBN 13: 978-1-63081-146-4

10 9 8 7 6 5 4 3 2 1

For my mother, Ioana
For my wife, Andreea

Contents

Preface

I wrote this book to tell you about a new and exciting new profession: the information technology (IT) architect. It is a great job responsible for coming up with the solution at the business, application, data, and infrastructure levels or even the enterprise level and follows up with its implementation.

Actually, this profession is not that new. It has been on the market for at least 30 years as it began gaining traction in the mid-1980s with the introduction of the Zachman framework. Why write another IT architecture book when the market has a lot of other books describing the various architecture frameworks and methodologies? After all, it should be quite simple to become an IT architect: you get a couple of books, you read them, and you start drawing some diagrams. Or is it?

When I became an architect more than 20 years ago I was attracted by a very interesting description from the *Sun Certified Enterprise Architect for J2EE Technology Study Guide* of an architect that I can remember even after all these years:

> The designer is concerned with what happens when a user presses a
> button and the architect is concerned with what happens when ten
> thousand users press a button.

As a developer, I created a lot of buttons using various programming languages, but I rarely had to think about performing the action associated with the button when different users pressed the button at the same time. The architect is the one concerned with the *performance* of the application handling the action and its ability to handle future load. Later on, I understood that other special requirements (also called nonfunctional as they are not related to the functionality of the application) like availability, scalability, and maintainability are also the responsibility of the architect. This was exciting stuff and something that really piqued my interest. I started to read more books about the IT architecture and what exactly I needed to learn to be able to perform this complicated job. Although there were quite a few books describing frameworks and methodologies for different kinds of architecture, there was no clear definition for each kind. One of the popular ones out there and quite close to my background was the application architecture. It seemed that being able to model applications or systems was an important skill, so I started to learn more about Unified Modeling Language (UML), which was quite a popular modeling language with clear definitions that helped communicate clearly the design of the applications.

Over the years, I clearly understood what the architect role was and got the knowledge required by working out of role as an architect, which helped me to get my first architect job later on. The journey was not an easy one. I had to struggle with the lack of clear definitions for the architect roles, with the lack of a guidance book that explained the requirements for the role, and, on top of that, with the inconsistencies across the IT industry with regard to the responsibilities for the role (which still remains an issue nowadays). On the hiring side, there was a lack of interviewing abilities as management did not have a clear idea of what they were looking for. If they were hiring an application architect, they were actually looking for a super-developer who would be able to code but also draw nice pictures for the stakeholders. If the position was called enterprise architect, the expectation was to draw nice enterprise pictures and put together powerful presentations that were presented to the executives to assure them the company was on the right track to achieve its business goals and strategies.

I wrote this book so that others get a smooth, or at least a less bumpy, road than the one I had and so that the companies can get the right people for the architect job and the IT practitioners who did their jobs for a while and want to grow can get a clear path of what they can become and a checklist of what they should do to get there. I wrote this book for developers who would like to understand what happens when 1,000 users press the button and how their application should be created to handle the load.

Audience of the Book

The book is written with three groups of people in mind:

1. *IT practitioners in roles other than architects including but not limited to:* software developers, programmer analyst, database administrators, business analysts, infrastructure specialists, applications engi-

neers, system analyst, system designer, software engineers, and system administrators.

Chapters 2 to 7 will give a clear understanding of the various architect roles, what the benefits of becoming an architect are, and the different paths that the IT practitioner should follow to become a successful architect. Both the technical knowledge as well as the performance competencies that will put them in front and will ensure they will get hired are discussed. There will be no more guessing what the required backgrounds for different architect types are and what the performance competencies are that were required to lead the project teams from the technical point of view. The architects will be able to communicate effectively with the stakeholders and influence them towards the implementation of the best solutions.

Chapter 9 will give the IT practitioners clear guidance on how to get the first architect job and what the employers are looking for in the successful candidate. The knowledge you will get out of this book will definitely put you on the right path to obtain that long-desired architect job.

2. *Existing architects:* Including but not limited to business, network, solution, system, infrastructure, data, security, and enterprise architects (for a complete list, please refer to Chapter 2). You might ask yourself why you need this book if you already an architect. What else can this book teach you that you do not already know.

The book is trying to help you successfully transition between the different architect jobs and to make you a more successful architect. Let us say, for example, that you are a solution architect, you have been a successful one for a number of years, and you would like to become an enterprise architect. This book is giving you the path to follow to become one. This is not a book about what certifications to take and what to read to get the certification (although I do mention some certifications that might make you more marketable). This book is giving you a checklist of things to learn on both the technical side and performance skills that will assure that you will be ready to transition to the new architect job. If you need to become a better leader (required skill as an architect), there is a section in this book that discusses the aspects of the leadership on which you need to focus as an architect, as well as the negotiation, presentation, communication, and other performance skills that you need on a daily basis.

If you are an architect but you feel that you might lack something (maybe you cannot present your solutions too clearly or maybe you are not able to influence the stakeholders and convince them that the options you picked are the right ones), then this book will help you understand what skills you are lacking and how to develop them to become a more successful architect without having a lot of expensive training courses.

3. *Management/executives who have architects as direct reports or plan to hire some.* As a manager, you might have an architect background (ideal case), so you know what to look for in a candidate, or you might not know a lot about the architecture, so this book may help you to understand the different architect types and how they bring value to the organization. Once you have a clear idea about which architect role you would like to have in your team, you can obtain a lot of information about the responsibilities and how to ensure that you hire the right people for the role, avoiding the painful experience of getting somebody on board only to find out later on that this person is not right for the job.

What Is in This Book

I already gave you a hint about the content of the book when we discussed the audience groups and how each group can use different sections of the book to get the most benefit. This book starts by quickly defining the various types of IT architecture in the industry and summarizing what the various architect jobs are and why becoming an architect is such a rewarding experience.

This book continues with the background required for an IT professional to become an architect: the technical background that focuses on the T-shape personality and the required ability for an architect to combine the deep expertise in key areas with broad experience and knowledge and the performance competencies (such as leadership and communication) that represent the key to being a successful IT architect. This book explores in detail the many ways to become an architect considering that different backgrounds allow you to become a different type of architect. It gets into details about the typical deliverables for an IT architect, their structure, and how to approach their creation.

This book focuses then on the benefits of architecture for the organization and how you can approach the successful creation (or the enhancement) of the architecture practice in a structured way. Once you have a good understanding of the background, the path to become an architect and how you can maximize the benefits of having an established architecture practice for the organization this book gives you the help you will need to actually get the job focusing on how you sell yourself and explain the value you can bring to the organization.

This book then wraps up with a discussion on the future of the IT architecture, including the current constraints and drivers for this change and how the required skills will evolve. It also includes a glossary of terms usually used in IT architecture and a comprehensive bibliography that will help you delve into different important subjects to prepare yourself for obtaining your desired architect job or simply becoming a better architect.

Chapter 1: Introduction

This chapter provides the necessary introduction of the IT architecture, its roots, and how it became clear that the IT industry needs another role apart from the existing ones. It describes what the objectives for the book are at a high level.

Chapter 2: What Is an IT Architect?

This chapter starts with a little bit of history and the origin of the word architect. It continues with the definition of the major types of IT architect, focusing on the major differentiators in between the architect and other IT roles.

Chapter 3: Why Becoming an Architect Is a Great Job

Having seen what the IT architecture actually is, this chapter discusses why an IT professional would like to become an IT architect. Different reasons are explored, including job satisfaction, financial aspect, and job security.

Chapter 4: A Skills and Competencies Profile for an IT Architect

After getting a better understanding of what an IT architect is and seeing the benefits of becoming one, it is time to focus on the background of the IT architect. In Chapter 4, I explore in detail what kind of general and technical skills and competencies are required for an IT architect. This chapter provides guidelines of how an IT practitioner has to prepare to become an IT architect. For all the skills and competencies, I indicate the proficiency levels that would differentiate the junior architect from a senior one. It starts with the common technical knowledge and continues with providing comprehensive information on the technical profile expected for an IT professional to be successful in an architect job.

Chapter 5: The "Soft" Background of an IT Architect

This chapter includes all the performance competencies (soft skills) required to become a successful architect such as communication, leadership, presentation, planning, consulting, facilitation, change and stakeholder management, emotional intelligence, and, one of the most important, being able to sell the solution to the stakeholders so that it will be successfully implemented.

Chapter 6: How to Start: The Road to Becoming an Architect

This chapter is the core section of the book, providing detailed guidance on how to become an IT architect, discussing each of the different paths that you might take to become an architect, and the depth that you will need to acquire along the way. The main dimension that provides the separation between other IT roles and the architect one is the level of detail. It discusses the evolution path to become an architect starting as a developer, analyst, infrastructure specialist, and database administrator. It provides the architecture specific knowledge including the drivers behind the evolution of architecture (Clinger-Cohen Act, Sarbanes-Oxley Act, and so forth).

Chapter 7: Architecture-Specific Knowledge

This chapter continues the discovery of the road to become an IT architect, focusing on the architecture-specific knowledge, including the architecture frameworks and methodologies specific to different types of architecture, architecture modeling languages, and so on. I discuss in detail the various architecture frameworks and methodologies for the different types of architecture as well as the most common architecture modeling languages.

Chapter 8: The Deliverables Produced by an Architect

This chapter focuses on the types of documents that are commonly produced by the various IT architect roles, giving specific examples for each of the sections and providing guidance on how to obtain information and structure the document to have the biggest impact on the audience. It also discusses the most common inputs into the architecture documents as well as how the architect is supposed to extract the information and use it as an input into architectural thinking.

Chapter 9: How the Architecture Brings Value to the Organization

This chapter will examine the architect job from the point of view of the employer, the manager, director, or executive who is in the position to hire. The question that every employer would ask is: why should they spend money from their budget to hire an architect? As we saw from the previous chapters, the cost of an architect is not trivial and usually it is not easy to convince your boss that you need to get a new expensive resource (that will not write code). We will look at this from the perspective of the person to whom the architect would report, focusing on what is the reason that a company needs architects and how the management can make sure that they will hire the best person for the job.

Chapter 10: How to Get the Job

If you got this far into the book, then it means you have an idea of what it means to be an architect, what kind of work you are supposed to do, and what education you need to get. This chapter assumes that you did your homework and used the guidance provided to embark on the road to becoming an architect. You read some of the books indicated in this book's bibliography, you started to apply some of the concepts in your day-to-day work, and you indicated to your boss in one-on-one meetings that your long-term goal is to become an architect, leveraging your experience and using the technical and business knowledge you acquired so far. This chapter will ensure that you know what role would be the best fit for you avoiding the painful experience of applying to the wrong role or being unprepared for the interview and the job.

Chapter 11: The Future of IT Architecture

The architect job is a relatively new one on the market. As with any other jobs, it will continue to evolve, requiring new skills and competencies or refining the existing ones. This chapter discusses in depth what new or modified architect roles might appear on the market and what new or modified skills these jobs might require. The evolution of the IT architect career is also constrained by the evolution of the organization, IT, and business but also by the standardization of the responsibilities for each of the roles. The architect jobs can be in the future commoditized across various companies instead of being specific to the business and IT departments within the company. By looking at the most important drivers, we will be able to take a glimpse into the future of the IT architect roles.

Glossary

This section summarizes the most used architectural terms for your future reference. It should be used as a central place to look when in doubt to get definition of the most common architectural terms.

Bibliography

This section provides a comprehensive bibliography to help you build on both the technical knowledge and the "soft skills" that you will need to transition to the architect roles.

How to Contact Me

I put a lot of effort in gathering this information, and I will gladly welcome any comments or feedback you might have; please direct them to becomeITArchitect@gmail.com.

Acknowledgments

Like most books, *How to Become an IT Architect* has been a long time in the making. The idea of providing a comprehensive guidance on how to become an IT architect (or improve your skills required for the job) dates back years ago when I realized that despite all the books discussing the architecture frameworks, patterns, and paradigms, there had been no publication giving a complete picture of the IT architect role. Most of the time, the soft skills that are essential to be able to sell solutions communicate effectively and create powerful presentations were overlooked.

I especially thank my family (my mother Ioana, my wife Andreea, and my son Razvan) for their support, which allowed me to write this book. Without their encouragement and understanding, I could never have completed it and I would have not being able to go through the long evening to late-night working hours.

A great debt in creating this book goes to the reviewers Eduard Telescu and Ed Tittel and to Artech House, whose editors, including Aileen Storry, Stephen Solomon, and Molly Klemarczyk, were very helpful and knowledgeable throughout the process of developing this book, making sure that my ideas stayed on track and were easy to follow.

I gratefully acknowledge The Open Group for permission to incorporate extracts of its copyrighted material from the TOGAF© 9.1 standard. During the years, the framework has been my guide through the tumultuous architecture oceans. I also gratefully acknowledge Marc Lankhorst and the book *Enterprise Architecture at Work: Modelling, Communication and Analysis* that showed me how to use the Archimate modeling language to build powerful models instead of nice diagrams.

1

Introduction

If you will indulge a trip down memory lane, I still remember how excited I was when I got my first Commodore 128, the most versatile 8-bit computer. Right then, the world of technology started to be reshaped by every human who now was able to interact with these machines through something called a programming language. I was in my first year at university ready to absorb all the knowledge that the new discipline of computer science had to offer. Looking back at those beginning years (Capability Maturity Model level 0) in the information technology (IT) industry, when things were done in an entirely ad hoc fashion and companies relied on the help of the heroes who had to put in long hours to make sure the implementations were successful, I wonder how we managed to accomplish as much as we did.

Eventually, those companies began to understand that developing entire systems and applications meant more than just making sure they had the best coding practices. The developers could not (and were not trained to) understand the implications of the information exchange between applications. When I started to get interested in IT architecture, I remember reading in a book how the developer is only responsible for what happens when one user clicks on a button, but the architect is responsible for what happens when a thousand users click on a button. This is a very simplistic way to explain why the architect's role is so important in the enterprise and why he or she gets paid so well.

One can wonder why this new role was introduced. After all, we had the developer/programmer who was able to design the applications and the system/business analyst who could elicit the necessary requirements. Adding an architect just meant more overhead, or did it?

Most companies saw the problem this way at the beginning: they thought that as long as they had some good developers capable of delivering good code, they were OK. I remember working for a Canadian bank quite a few years back and developing code for a wealth management application. Everything was going well, everybody was doing their part, and the project seemed to go along just fine until a couple of weeks before the implementation when it all started going sideways. We finally managed to integrate the various components and, from the functional point of view, the system was doing what was expected. However, when business users started to test it, everything changed for the worse. The system was responsive to a point, but then it started to run very slowly. Eventually, it crashed. As developers, we checked the code and everything seemed to be OK, but because the issue was quite serious, we started to look into how various application components interacted and tried to figure out where the latency was coming from.

This might be called the infancy of Enterprise Java: the concepts were fine, but our practices were not. As this was way before automated tools like Mercury LoadRunner appeared on the market, we had to figure out what was wrong manually. We had to put in a lot of long hours reading the logs and trying to find out what happened when more than a few users tried to perform the same operation. As this was at the beginning of the distributed, n-tier architecture, it meant that we had to focus primarily on the application server and connectivity to the database. In the end, the issue was all about the complete lack of a high-level view of how the various application components interacted. Everybody cared about the functionality, requirements-wise, but nobody cared enough about performance and its requirements to do anything about them.

This was one of the first times that I was in a specific situation where the need for an architect became evident. The myth of the developer who could code, debug, and save the application was dead. The management of the company started to realize that they needed a new role on the development team, someone who could oversee the whole structure.

One thing was obvious: he or she should know a lot about the application, but at what level of detail? Ultimately, everybody has limited memory capacity and equally limited time. No one person can possibly know and own every little line of code that composes a complex application. When we expand that picture to the whole enterprise, then we quickly understand that unless we talk about a super-human, there is no way that an architect would know everything about all applications. What is the solution? This is the moment when the concept of T-shaped skills was introduced by David Guest in 1991 [1].

The concept of knowing just one field/domain but being able to use summarized information from other subject matter experts to provide a solution that would span multiple fields/domains was revolutionary. Years ago, there was no clear idea about what skills would be required in an IT architect and, most importantly, how an architect could add value to the company. However, when comparing success stories of those employees who had adapted well to the role, the stories of those who had not, there was one distinguishing factor.

That thing became obvious in retrospect. Although technical knowledge (including the IT architectural design and description) was necessary, one of the most important things was the ability to communicate, lead, and influence other people. These all fall in the realm of soft skills, outside of purely technical matters. Why communication and leadership? Because without those two skills, a technical guru was not able to communicate his or her solution to the developers and other members of the team and he or she was not able to lead them to implement it. Successful IT architects were able to get buy-in not only from stakeholders, but also from the team and to break the barriers between themselves and the implementation staff. They were not labeled "ivory tower" guys who would work in an isolated office producing dreamy architectures detached from IT, business, and budgetary realities. Their architectures would not only offer solutions, but would also provide the best fit considering all constraints and assumptions.

Then another (new) dimension was added to this mix. The ultimate goal of any company is to make a profit, and this cannot be done without a clear business vision, mission, and values. Another type of IT architecture tried to link business and IT and to show that ultimately the value of IT architecture is to maximize the business value by choosing the right solution at the enterprise level encompassing both business and IT concerns. Naturally, this role was called the Enterprise Architecture. Imagine one has somebody (called the Enterprise Architect) who would be able to understand the business architecture as well as the IT architecture and choose the right solution at the enterprise level that would allow the company to be more competitive, perform better than its peers, and ultimately make good profits for shareholders. The purpose of this book is to sell you on the idea of becoming an IT architect (whatever flavor might be appealing to you) and somehow transmit the passion for becoming an IT demigod. That is why this book starts first by exploring what might motivate someone to become an IT architect.

This book also provides a definition of the various kinds of IT architects that exist today including the skills and knowledge needed for each such role. I review each of them in detail and provide resources about where you can get more information and smooth your path to become an IT architect. As with any new job, there are a lot of people who would like to know more to help them decide if this is a career path worth embarking on before they invest a lot of time, money, and energy into getting the knowledge and experience required to secure an IT architect role. This book also provides information about the knowledge required to become an IT architect including the architecture frameworks, modeling languages, views and viewpoints, and more. Going into detail on these frameworks and languages is outside the scope of this book, but you will get an idea of what do you have to study to make sure you have the necessary technical skills and knowledge to become an IT architect. At the end of the book, I will also share some techniques on how to get a job after you have the necessary knowledge and skills to fill an IT architect position.

In a nutshell, the purpose of this book is not to sell the architecture concepts and explain why every company bigger than a few people should have at least one IT architect, but to sell the idea of becoming an IT architect and to provide some guidelines on how to become one. I will give you a checklist of what you should do to become one. For more information about the various kinds of roles one might fill as an architect, please consult the bibliography at the end of this book.

I hope you will enjoy this book and that I can manage to convey the strong passion I have for this role. I hope that, after reading it over, you will want to embark on the path to become an IT architect and provide architectural solutions not just for the present day but for years to come.

Reference

[1]   Guest, D., "The Hunt Is on for the Renaissance Man of Computing," *The Independent*, September 17, 1991.

2

What Is an IT Architect?

It would be useful to start this discussion with a definition for an information technology (IT) architect. Unfortunately, defining the term IT architect is not an easy job for at least two reasons:

1. Architects have had an established role in construction for centuries, and the term architect is assumed by most to refer to a building architect rather than an IT architect.
2. There are so many flavors of the architect job in the IT world because of the lack of standardization.

For the purposes of our discussion, I will refer to the IT architect role to cover any of the types of architect roles currently understood and sought-after in the marketplace.

Let us begin our discussion with the word "architect." The roots of this word come from the Greek "arkhitekton," which means chief builder [arkhi ("chief") + tekton ("builder")]. By extension then, an IT architect would be the chief builder for IT systems rather than buildings or other physical structures.

An IT architect is the person who comes up with a technical solution that utilizes IT resources to meet specific business requirements. This is an important clarification because IT practitioners sometimes forget that the only reason IT exists is to serve the business. The IT architects determine which information technology investments will yield the best profits (return on investment), both in terms of hard costs and productivity benefits.

IT architects strive to bring operational efficiency to an organization through information integration and management. We will get into more details later, but most of the architecture frameworks focus on the business value that a certain project or initiative provides to an organization. One can prioritize these project/initiatives be more effectively by comparing the business value that each one of them provides.

In general, the term "architect" refers to a person who can come up with a high-level solution for an application, a system, or the entire enterprise (that is why there are different kinds of architects). Once a solution is chosen, the architect can also translate that solution to fit the individual viewpoints for various stakeholders who may need to buy into that solution before it can be implemented.

2.1   Different Kinds of Architecture

Defining architecture is not as straightforward as defining other types of disciplines in the IT world, mostly because architecture covers so many possible forms. Apart from the actual architectural domains that we discuss later, there is the enterprise architecture. A person in this role tries to encompass both the business and the IT domains. He or she has as his or her main purpose harmonizing the two to the point of creating information systems that not only satisfy the business needs, but are flexible enough to accept changes. So I guess the best way to start would be to talk a little bit about each of the forms that IT architecture can take.

2.1.1   Enterprise Architecture

The definition for enterprise architecture, according to Gartner [1], is:

> Enterprise architecture (EA) is a discipline for proactively and holistically leading enterprise responses to disruptive forces by identifying and analyzing the execution of change toward desired business vision and outcomes. EA delivers value by presenting business and IT leaders with signature-ready recommendations for adjusting policies and projects to achieve target business outcomes that capitalize on relevant business disruptions.

The MIT Center for Information Systems Research (MIT CISR) defines enterprise architecture as the specific aspects of a business that are under examination [2]:

> Enterprise architecture is the organizing logic for business processes and IT infrastructure reflecting the integration and standardization requirements of the company's operating model. The operating model is the desired state of business process integration and business process standardization for delivering goods and services to customers.

The main purpose of this emphasis on architecture is to improve the effectiveness and efficiency of the business itself by making sure each IT initiative or change comes with some business value attached.

An enterprise architect should be able to see the big picture and link the business strategy to specific technology capabilities. People always compare enterprise architecture with the plan for a city. Its main purpose is to align business and technology needs and deliver solutions that are independent of the technology platform and vendors.

Enterprise architecture includes a few domain architectures that I will discuss next.

2.1.2 Business Architecture

The main focus for a business architecture is to view and understand the core business processes, functions, roles, and capabilities, as well as the operational needs of the enterprise. An architect would collect input information including the stakeholders' concerns, business principles, goals, and strategic drivers. Based on these, the architect will document the current and target states as well as the architectural gap and a business roadmap. Along the way, the architect should use the most appropriate viewpoints to show how the already-mentioned concerns will be satisfied by the target state.

An architect should ensure that the architecture is flexible enough to accommodate any changes in the business strategies and vision. Depending on existing documentation for the business architecture, an architect:

- Might only need to update it with the new or modified business process and functions;
- If no documentation exists, will need to document, verify, and get the buy-in from the key stakeholders for the business processes and functions that the architecture should support.

Business architecture is usually a prerequisite for other architectural domains. Having identified the business processes, the next question is which ones would be done manually and which ones require automation and therefore will be implemented by an application or service.

2.1.3 Information Architecture

Information architecture focuses on how the enterprise can organize and make use of information at the enterprise level to support the business processes, functions, and roles defined in the business architecture. Because we currently live in an information age, having the information needs properly architected is essential to creating a competitive advantage to manage, organize and access that information.

This goes back to the boundaryless information flow problem as defined by Jack Welch when he worked for General Electric [3]:

> It is essentially the problem of getting information to the right people at the right time in a secure, reliable manner, in order to support the operations that are core to the extended enterprise.

This is basically the ultimate goal for the information architecture. That is because the architect is the person required to make sure that the data is organized in the right way and that all the right people can access it.

2.1.4 Application Architecture

Application architecture is a second subdomain within enterprise architecture. It focuses on how the business services and functions are supported or automated by application services that are ultimately implemented by application components.

These applications may be described as logical groups of capabilities that manage the data objects and support the business functions in the business architecture. Basically, this type of architecture describes the components, services, and application layers that comprise the software design for the enterprise. This architecture should provide the support required to gain a better understanding of the detailed application-level design of the services and components to enable coding to begin. In the past, there were quite a few architectural styles for applications, including:

- Client-server: The system is separated into a client and a server application where the client makes a request to the server.

- Layered/tiered architecture: Functional concerns are separated into stacked layers. In the case of a tiered architecture, each tier usually resides on a different machine.

- Message bus: Software acts as a mediator between systems with different interfaces, so that the applications/systems can intercommunicate without knowing details such as message format and external data representations.

- Object oriented: Tasks for an application or system are divided into individual reusable and self-sufficient objects, each containing the data and the behavior relevant to itself.

- Service-oriented architecture: An architectural style (or paradigm) supports service orientation as a way of thinking in terms of services (self-contained, black box, logical representation of a repeatable activity) and their interfaces.

Based on the architectural requirements, constraints and business objectives, the application architect has to choose either a commercial-off-the-shelf (COTS) application or architect an application or system from scratch taking into consideration the existing architectural landscape. In the case of a COTS application, the architect takes an additional risk because the vendor may go out of business or decide not to support existing components, basically throwing away the entire effort to implement/integrate the application.

2.1.5 Technology Architecture

Another subdomain within enterprise architecture, technology architecture, focuses on the infrastructure services required to support the various application components. A specific application might need a message queue to asynchronously transfer messages between a message producer and consumer. Alternatively, it might need a secure file transfer service to ensure secure and efficient transfer of files using various specialized protocols. These services will ultimately be implemented on separate physical or virtual machines. The technology architecture also includes any kind of hardware required to support the application components identified in the application architecture including firewalls, load balancers, and servers.

Figure 2.1 is an example technology architecture diagram that illustrates some important technology components including:

- Router: A device that forwards data packets across the networks;

- Firewall: Security hardware or software system that controls the incoming and outgoing network traffic based on a set of rules;

- Virtual server: A server that shares hardware and hardware and software resources with other virtual servers;

- Physical server: A physical machine that can be partitioned into one or more virtual servers to maximize resources and allow use by multiple tenants;

- Hypervisor: A fundamental part of virtualization/cloud technology used to generate virtual server instances of a physical server;

- Virtual infrastructure manager (VIM): A manager that manages IT resources and relies on a centralized management module.

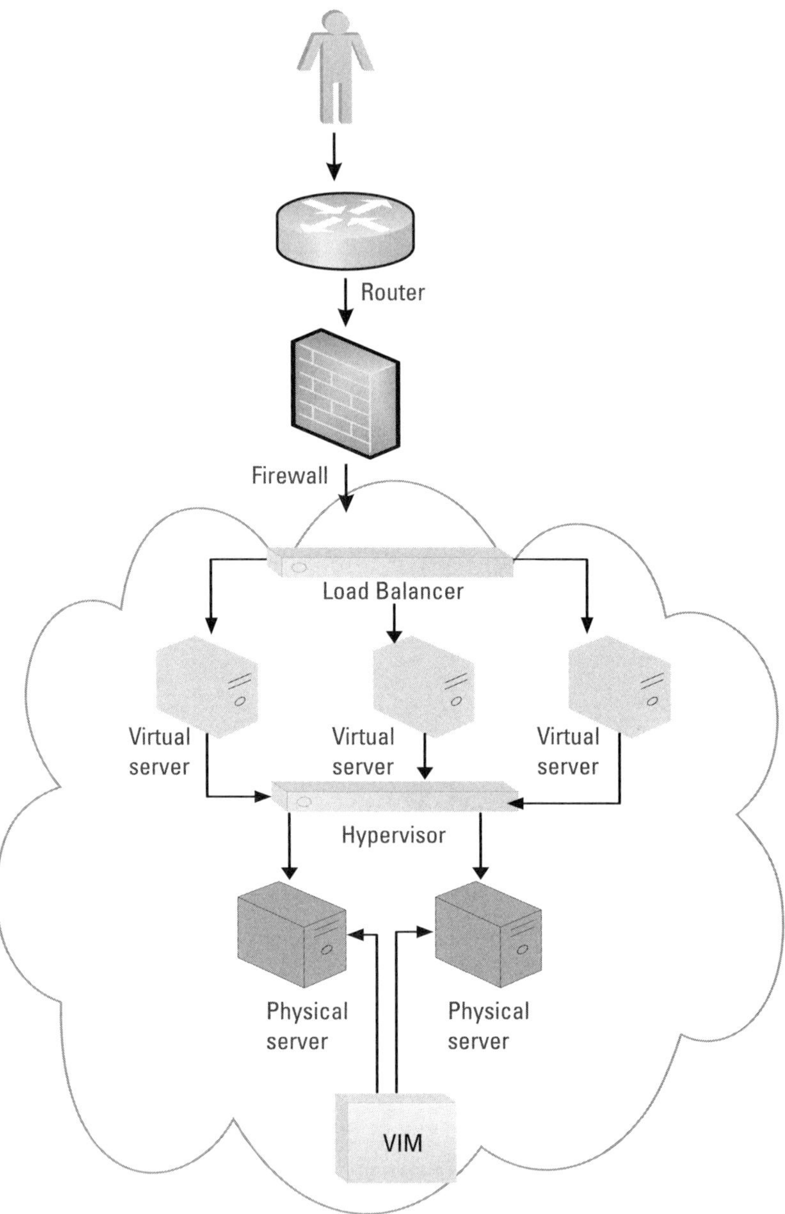

Figure 2.1 Technology architecture diagram example.

The architect will have to document the technology architecture and decide on the vendor product selection based on the existing architectural standards, policies and procedures. He or she should keep in mind that reusing the current products (especially where incremental capacity is added) must always be the preferred option.

2.1.6 Hardware Architecture

Simply another term for infrastructure or technical architecture, hardware architecture defines the components, services, and infrastructure layers that comprise the hardware high-level design. Mostly this means documenting the topology of the hardware resources and the deployment of the infrastructure services to support the hardware procurement processes and vendor selection as well as supporting the implementation. A hardware architecture diagram usually depicts the client and server nodes in the hardware configuration, along with infrastructure services and applications as well as the protocols and networks that interconnect them.

Like other domain architectures, infrastructure architecture is all about applying patterns and specialized mechanisms to solve infrastructure problems using well-established and proven solutions. Nowadays, there is increasingly more required from the infrastructure than ever before. The infrastructure is tasked with handling tremendous amount of data and facilitates the use of mobile equipment. For example, an emerging and increasingly popular cloud infrastructure architecture involves the use of different mechanisms such as a hypervisor (to generate virtual server instances on a physical server), failover system (used to create the reliability and availability of the resources by providing redundant implementations), and pay-per-use and audit monitors (to measure usage of the resources for billing purposes and collect data for audit purposes, respectively). By combining these mechanisms, an infrastructure architect can put together solution building blocks that can be used for the various problems that infrastructure architecture is trying to solve. The architect needs to be able to select the most appropriate viewpoint to illustrate how the solution that he or she proposes meets its business and technical requirements. Different architecture frameworks and modeling languages have different viewpoints from which the architect can select such as:

- Infrastructure usage viewpoint: To illustrate how the applications are supported by the hardware infrastructure, helping with the analysis of the performance and scalability.

- Implementation and deployment viewpoint: To illustrate how one application is realized and deployed on the infrastructure.

- Technology standards catalog: To document the agreed standards that are currently in place and that will be implemented/decommissioned in the near future.

- Application technology matrix: To document how the applications map to the technology standards or technology platforms.

2.1.7 Solution Architecture

Solution architecture aims to address specific business problems and requirements that span multiple technology domains. A solution architect develops solutions that fit within the enterprise architecture in terms of its systems portfolios, integration requirements, and so forth. Solution architects usually take a deep tactical approach to a limited amount of applications and systems in contrast with enterprise architects. The latter take a more strategic approach, so as to optimize the development and communication of a large number of systems/applications according to the business strategies and directions on how to evolve the business capabilities. The architectural solutions proposed have to take the following items into consideration:

- Business requirements that define the required functionality, usually obtained from the client and facilitated by the business (or systems) analyst;

- Project or program delivery constraints that combine the direction from the enterprise architecture for the affected systems/applications with the directions from the account/program and project managers that usually introduce time constraints;

- Existing architectural landscape; usually the solution has to work in an existing architectural landscape and may have to reuse existing vendor products, services, and so forth.

2.1.8 Systems Architecture

Systems architecture focuses on those components that make up the system and how they are linked and integrated with one another to accomplish the system's functionality. A system architecture approach focuses on the system as a whole (including the application, data, and infrastructure domains) linking what is desired (requirements) with what is feasible. One of the major concerns for a systems architect is the system interface and how it interacts with other systems. To describe the systems architecture of an IT system, you need to describe at least the application data as well as the infrastructure architecture. An architect needs to select the most appropriate views to illustrate how the components work with each other and how the various systems interact at different domain architecture levels. To select the components that will make up the system, the architect needs to have a thorough understanding of the business capabilities and processes that the system must (or will) support. This architecture needs to depict the current state and project target state as well as its end state using different views that include:

- Service realization viewpoint showing how the various application and infrastructure services are realized by application or infrastructure components;

- Application structure and application behavior viewpoints showing the structure and internal behavior of the application/system components (see Figure 2.2);

- Infrastructure viewpoint showing the hardware, network, and software products upon which the application layer depends to perform its functionality (see Figure 2.3).

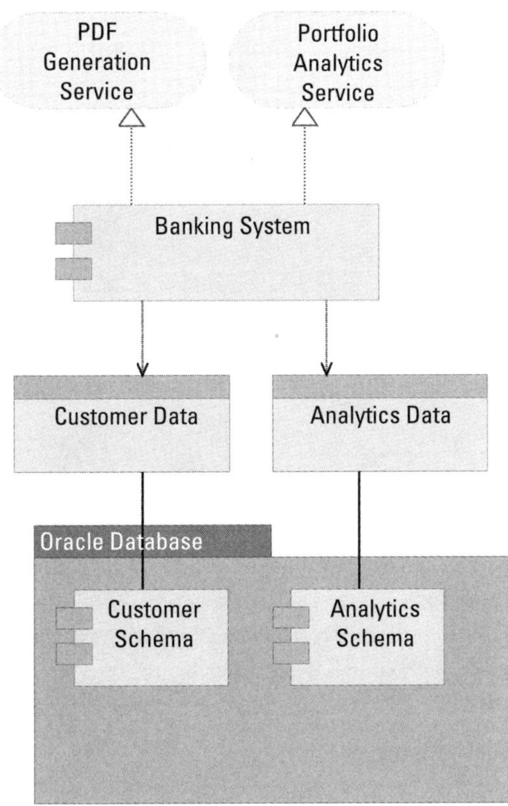

Figure 2.2 Example of an application structure viewpoint.

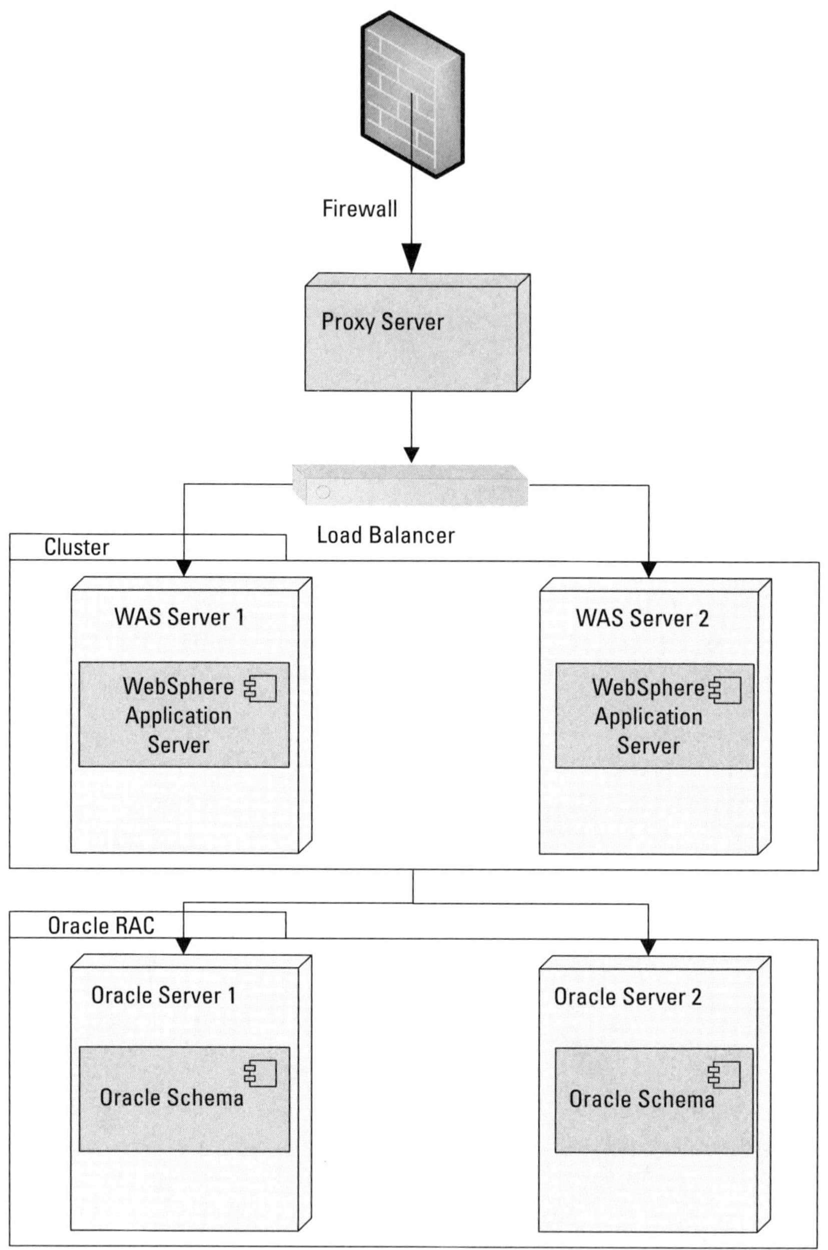

Figure 2.3   Example of an infrastructure usage viewpoint.

2.1.9 Network Architecture

Network architecture provides a complete picture of the logical and physical network components including hardware components used for communication (firewalls, routers, and so forth), network layouts and topologies, physical and wireless connections, and cabling as well as the protocols used. The diagrams used for this kind of architecture might include the following:

- Different network types such as wide area network (WAN), metropolitan area network (MAN), local area network (LAN), and virtual private network (VPN);

- Different network devices such as switches, routers, firewalls, wireless access points, and WAN optimization devices;

- Different network technologies such as Ethernet, carrier Ethernet, and Synchronous Optical Network (SONET);

- Different network layers and the protocols used at different layers starting from the application layer and going down to the physical layer [for more information, please refer to the Open Systems Interconnection (OSI) network model developed by the International Organization for Standardization] [4].

Network architecture, like any other domain architecture, is based on a few principles such as layering—the division of the communication tasks into smaller parts, each part accomplishing a particular subtask.

2.1.10 Security Architecture

Security architecture covers all the processes and mechanisms by which computer-based equipment, information, and services may be protected from unintended or unauthorized access, change, loss, or destruction. This includes not only protection from unauthorized activities or untrustworthy individuals but also from unplanned events and natural disasters. Security architects focus on the security measures that are required to protect the enterprise from threats while making sure day-to-day tasks can be conducted. This implies securing corporate networks, and their core systems, and also defining an appropriate security-level classification for information technology assets and the services model in an IT security infrastructure. Security architecture includes but is not limited to authentication, authorization, and encryption, protection from malicious software, auditing, and control processes.

To define the security architecture, the enterprise must adopt guiding principles. The most common such security principles include:

- *Least privilege:* Each user or system component should be allocated sufficient privileges to accomplish its specified functions, but no more.

- *Deny by default:* That which is not expressly allowed is denied.

- *Secure failure ("defense in depth"):* A failure in a single system function or mechanism should not lead to violation of security policy.
- *Continuous protection of information:* Information must be protected to a level consistent with its security classification regardless of the processing context or environment (e.g., storage, transmission, development phase).

The enterprise might also decide to develop a reference architecture that will include a security model, thereby providing a standardized approach to define the level of security appropriate for any given solution design or initiative. Such a model includes technical capabilities that would help select the security technology standards and decide the most appropriate security architecture for any given application, system, or network. Enterprise assets security can be also classified using:

- *Sensitivity:* Different levels of the sensitivity of the information such as public, internal, confidential, and restricted;
- *Criticality:* What is the impact on the business if the information is altered or corrupted?

Security architecture is also responsible for defining the network zones (usually based on the principles specified above) that have different levels of security. The most simple network zoning consists of placing a demilitarized zone (DMZ) between the Internet and the internal network containing core systems. The outer firewall allows a significant variety of traffic into and out of the services in the DMZ, but little or typically no traffic is permitted between the inner region and the Internet. Nowadays, this is considered a very simplistic approach used to illustrate the defense-in-depth principle. More complex layouts are possible and, in many cases, appropriate, but most are some derivative of this form, with layers of protection wrapped around one another.

2.1.11 Other Architecture

IT architecture is continually evolving and keeps adding more architecture types to the already existing collection. The four main domain architectures (business, application, data, and technology) are also split into subdomains and some architectures are horizontal spanning across multiple domain architectures (e.g., security architecture). The lack of a universally accepted definition for each of these architectures, together with the truism that each enterprise has its own architecture types and definitions for the responsibilities assigned to certain types of architect only increases the confusion.

Looking online reveals numerous job descriptions for solution architects that include in their responsibilities' code review. Also, enterprise architects require in-depth knowledge of certain technologies up to and including configuring application servers.

A few other types of architecture that may be encountered online include:

- *Cloud architecture:* The establishment of the well-defined architecture solutions composed of combinations of cloud computing mechanisms (combinations of specialized IT resources that act like solution building blocks for cloud architecture) together with the interactions between them.

- *Messaging architecture:* Combination of messaging (enterprise integration) patterns used in complex messaging systems to ensure the loose coupling of the applications. This may also be encountered as messaging queuing architecture (an architecture for distributed systems based on the concept of reliable message queuing).

- *Digital (or mobile) architecture:* Specialization of application/software architecture with a focus on the development of digital solutions for mobile devices requiring a deeper understanding of mobile technologies and their relationship with the core enterprise systems.

There are also other types of IT architecture, but these ones are the main ones that can be found in cyberspace.

2.2　What Does It Mean to Be an Architect?

So far, we have discussed what different kinds of architecture exist, have offered a definition, and have provided some information about the responsibilities for each type of architect. As mentioned earlier, the purpose of this book is not to sell the architecture and explain why every company bigger than a few people should have at least one architect. Rather, the goal is to sell the idea of becoming an architect. Many good developers think about what is next for them after quite a long career as a developer. Most of them see becoming an architect as the next logical step, and some of them would be right to take that step. This book will help with that decision so you know upfront what you should look for, what kind of architect role is the best fit for you, and what nontechnical skills you need to develop to be successful as an architect.

What exactly does it mean to be an architect and how is this job different from the other types of professions in the IT world today? The answer is quite simple: the focus of the architect job is not on the implementation of the design but in coming up with a design that solves the initial problem. This design is called a high level-design to differentiate it from the normal design that a developer might produce. Based on the type of architecture developed, it might solve a business problem (and in this case show how the technical solution satisfies the business problem) or a technical problem (like how various applications would interface with each other or why the solution infrastructure is the best fit for the application). The design would typically show the various systems and how they interface with each other or in case of the enterprise architecture the business capabilities, functions, services, and roles and how they are realized by application services and components, which, in turn, are realized by infrastructure services and components.

None of the famous buildings in the world could have been constructed without an architect drawing up plans and designs for the building nor without having a vision of how it would look and function once it was built. Construction architecture has been around for a very long time. IT architecture as a concept is new primarily because IT itself is no more than four or five decades old. During most of this time, the emphasis lay on the development of the actual application. Little or no thought went into creating a high-level view of the application/system and how it interacts with the outside world. Even littler thought went into showing or measuring how the application itself contributes to realizing the business value.

Reading over the preceding lines, most techies would tend to say, "Hey, this sounds like a great fit for me, sign me up!" Before jumping into the architecture boat and embarking on the way to becoming a great architect, there is one more thing that you must know: the job of architect is also about convincing others that your solution is the most appropriate one and that it satisfies all their concerns. This sounds quite easy, but given human nature, it is not quite so simple or straightforward. As an architect, you will work with other subject matter experts (like lead developers and infrastructure specialists) who have a lot of experience with the technology, application, or system that you are architecting (or changing the architecture of). Most of the time, this means that they expect you to be even more knowledgeable than they are for them to accept your solution. Your job is to convince them that your architecture is only as good as the information you get from them and without their support you cannot come up with a good solution. A good architect knows how to first earn the respect and trust from the subject matter experts and from all the members of the project team. This is the first step toward leading such a team. Although you might not have the technical depth required to produce the solution architecture by yourself, you can get develop that in discussions with the subject matter experts (SMEs) if you meet these two conditions:

- The SMEs are willing to provide you with the necessary information. Sometimes SMEs may perceive you as a "new kid on the block" who will try to steal their job or at least steal the success of their work. That is because you will be responsible for most of the interactions with the stakeholders instead of them. You can overcome this by earning their trust. This takes time and real effort, but it can be done, and I will tell you how.

- Know how to do your job. Most of the time the architect's job is about knowing what questions to ask. In my opinion, this is the most important skill that the architect needs to develop and cultivate. We will discuss more about this in a later chapter, but knowing what kind of questions to ask might make the difference between good and exceptionally good architects. This applies not only to SMEs but also to all stakeholders. As their time is in very scarce supply, you only have a few shots to get their attention and extract the information you need to incorporate in your decision requests or architecture documents.

Your interaction with the SMEs will ensure that you select the most appropriate architectural option, but most of the time as the architect you will be in the privileged position to discuss the solution with management and executives (account managers, application delivery managers, and so forth). Their interest is much different from that of the SMEs and, as the architect, you have to present to them a different view that would satisfy their concerns. Most managers do not have any technical background, so the view that you will present to them needs to be a translation of the technical view into layman's terms that can be easily understood. In a way, the architect acts as the interface between the project teams (and SMEs) and the management/executives (see Figure 2.4). In the case of the enterprise architect, he or she needs to translate the architecture in business terms and show how the proposed solution is in line with the business vision, mission, and strategy as well as which business value the proposed solution will bring to the organization.

Therefore, it is not just about documenting the technical solution and getting the buy-in from the project team. It is also about translating solutions into business terms and selling them to business management. This chapter covers most of the scenarios and interactions between the architect and the outside world, although for many architects there is just the interaction with the technical team and minimal interaction with the account manager. Although one kind of architecture is no more important than any other, there is a classification of architectures based on the complexity and breath starting from the application architecture, which can be sometimes done by a lead developer who understands the relationship between the various application components

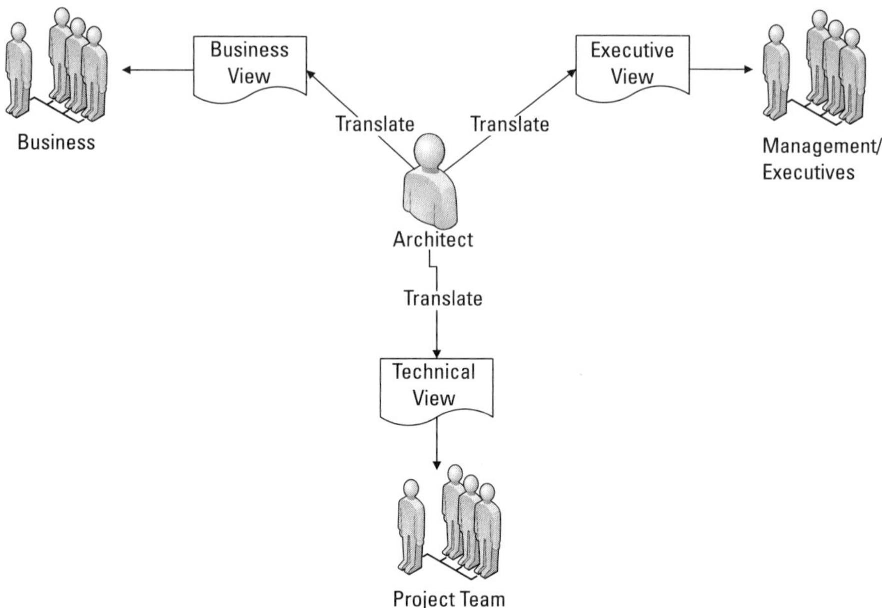

Figure 2.4 Architect as the interface between technical, management, and business teams.

and services and ending at the enterprise architecture, which needs a deep understanding of not only the IT world but also the business one. Nevertheless, at each of these levels of architecture, one of the most important things for architects is not their technical background (although they would not be able to perform his job without the technical base) but the approach that they use to collect the required information and convince the others why their solution is the best one given the constraints imposed.

References

[1] Enterprise Architecture (EA), IT Glossary, http://www.gartner.com/it-glossary/enterprise-architecture-ea/. Accessed May 2015.

[2] MIT Center for Information Systems Research. "Enterprise Architecture," http://cisr.mit.edu/research/research-overview/classic-topics/enterprise-architecture/. Accessed April 5, 2016.

[3] Integrated Information Infrastructure Reference Model TOGAF® Version 9.1, http://pubs.opengroup.org/architecture/togaf9-doc/arch/chap44.html. Accessed February 2016.

[4] The OSI Model's Seven Layers Defined and Functions Explained," https://support.microsoft.com/en-us/kb/103884. Accessed March 2016.

3

Why Becoming an Architect Is a Great Job

So far, we have discussed what kind of information technology (IT) architectures are out there, some of the surrounding history, and what it means to be an architect. This is a good start, but the decision to become an architect is also influenced by a number of other factors. Such factors include the most obvious ones: job satisfaction, financial rewards, and job stability. In this chapter, we will go through all of these aspects and analyze how the job compares with other similar positions in the IT industry. Now is the time to consider what is in it for you. Thus, what follows are some of the benefits that can accrue from choosing a career in IT architecture.

3.1 Job Satisfaction

All companies want to attract the best employees in the market: they are the ones who produce most efficiently and therefore make the company the most profitable. Often, employee engagement surveys include sections specifically written to find out what makes a great employee stick with a company and not look for another job elsewhere. One would expect that more money would attract the best people. As it happens, the financial aspect is not always the most important one. Although we are motivated by our income, money is not the only reason we get up early in the morning and go to the same job each and every day.

According to a study by MSW Research and Dale Carnegie Training [1], which explored key drivers for employee engagement, compensation (pay) was not among the top elements that drive such engagement or that influence a person's decision to leave a company. Instead, the study revealed the top three key drivers to be:

1. Relationship with the immediate supervisor;
2. Belief in senior leadership;
3. Pride in working for the company.

Other factors that also influence engagement and an employee's decision to stick with a company include:

1. Recognition for one's work;
2. Amount of influence that one has in decision-making;
3. Opportunities for career advancement;
4. Periodic evaluation and feedback;
5. Working conditions.

By now you may be asking yourself what this has to do with being an architect. This is a good question.

Let us talk about the first factor in the list above: what it means to be recognized for the work that one does as an architect. Such results and recognition might not be immediate. It may take months, or even years, for your efforts to bear fruit. When they do and there is a direct correlation between your efforts and the solution that was implemented or between the business value and expected results, then a good reputation will follow.

The second factor in the same list is perhaps the most important aspect of the architect's job: influence. More companies now understand the role and benefits of IT architecture and have started to trust the decisions proposed and documented by their architects. Depending on the type of architecture involved, there could be decisions that would affect the performance, reliability, or any of the nonfunctional properties of an application or system. There may be a decision that could shape the enterprise architecture and put your company ahead of your competitors. The more influence that you have on making important decisions for your team, department, or company, the more satisfaction you feel when this decision proves to be viable in the long term.

For architects, opportunities for advancement are almost a certainty. There are frequent opportunities to move from less complex architectures to more complicated ones. As an architect, this is your chance to display your leadership competency. Such occasions are useful if you are looking to be promoted to a management position, for example. The validation of these job satisfaction drivers can help to make the architect's job even more attractive.

3.2 Financial Aspect

Although compensation scores very low on the MSW Research and Dale Carnegie Training's list of job satisfaction drivers, we have to recognize that most people do not come to work just for the pleasure and satisfaction of doing their job, but because they need money. Most of us have mortgages, credit cards, and bills to pay on a monthly basis. For those who aspire to be an architect, we cannot say that the associated salary's importance is close to zero. This is why, after we satisfy our craving for recognition and influence, we ask ourselves what kind of pay we can expect from a job like this.

There are many factors that greatly influence the kind of salary you might receive, including:

1. *The type of architecture you are producing:* Although no architecture is considered more important than the others, there are differences in the salaries of the various architects. Discussing salary ranges is difficult, as they differ from country to country (or even state to state), and is also outside the scope of this section. In general, however, domain architects (business [2], application [3], data [4], and infrastructure) appear to be paid less than enterprise architects [5]. However, this is subject to variance, for example, if an enterprise architect lives in a low-paid region, he or she may earn less than an application architect living in a medium to high-paid region. However, if we compare like to like (same type of architecture, same type of industry, and same region), then the preceding statement holds true. In some cases, certain niches or new types of architects, like the business intelligence (BI) architect, might actually make more money because they are in higher demand. System architects [6] and solution architects [7] are next in terms of average salary because they span multiple domain architectures (application, data, and infrastructure).

2. *Full-time and contract:* This concept applies to all IT jobs because there is a large differential between full time salaries and contract ones. On average, contract work earns contractors more than their permanent-work counterparts [8], and in the IT industry, many contractors stay with the same company for 2 years or more. Nevertheless, there are a few things that you must consider:

   • A contract is not a full-time, salaried position .The company could terminate your contract at any time, especially if you are a very expensive resource. For example, an enterprise architect who was brought in for a specific engagement will likely be let go when the project is finished. If you are seen as an expensive resource, the company might not want to pay you for more than a few months. They might expect you to leave right away and coach the permanent resources to increase productivity after you leave.

- As a contractor, your reputation is very important. One wrong move or one assignment that did not end well, and you might have trouble getting another contract for a while. Also consider that it can be difficult to live without a salary while you build your reputation. As we will see in Chapter 4, building strong performance competencies, including the way that you negotiate the solution and sell it to the stakeholders, can make a big difference. Remember, the contractor sometimes needs to find better ways to sell his or her solution over those of the other project team members, or managers, without damaging egos.

- In many countries, a fixed-term contract means that you have no benefits: no medical or dental insurance, no paid pension plan, no employee stock purchase plan, or other benefits that permanent, salaried staff may have.

- With regard to level of seniority, as an architect, a company will not normally hire you unless you are in a senior position, someone who can "hit the ground running" from the first day you start. As with other IT contract jobs, you will not usually be trained by the company and will be expected to have the knowledge you need prior to starting the contract. Unless you have a very specialized, sought-after skill set, the company would probably prefer to hire someone with a wide breadth of knowledge and a lot of experience.

3. *How good you are:* As mentioned before, we will later delve into the required competencies, so that we can distinguish what makes a good architect. For example, as an enterprise architect, you may be the chief information officer's consultant for any problems that affect the overall architecture, requiring good interpersonal chemistry between you and the chief information officer. Should he or she leave that position, you may have to look for another job as well. For now, let us summarize some of these key qualities:

   - The first one is technical background. You cannot be a high-level advisor and decide on the architecture for the whole organization if you do not have the technical background to participate in discussions with the subject matter experts and gain their trust. (Keep in mind that they might think of you as just one more overhead for the project at first.)

- The next most important thing is to have the soft skills needed to convince the project team members, as well as the stakeholders, that your solution is the best one. Senior developers aspiring to become architects need to understand that persuading others to get the buy-in is a necessary skill. If stakeholders do not understand the solution, or do not agree with it, then the architect has to do a better job of explaining it and addressing any issues raised. It is very likely that there are extremely competent architects out there who do not make the highest salary that they could because they do not have commensurate abilities to communicate, provide leadership, or manage the time spent working on a project.

- Another important skill is managing your stakeholders. Make sure that you handle all the "political" situations effectively and understand the culture of the organization. Consider the concerns for each stakeholder and how you can come up with a solution that represents a good compromise or consensus among all of them.

- Last, but not least, a great architect has the ability to ask the right questions. Anybody with the right background can ask the right questions, but architects who do not know much about a specific system need to be able to ask for the required information from project team members so they can then design the right solution. This is not about facilitation (although a good architect should know how to do that too) but about knowing what questions to ask to get the information you need. This concept will be discussed further in Chapter 4.

3.3 Job Security

We hear and read a lot about job security and what it means to regular employees. Years ago, it seemed like job security was assured; most people would start their career with one company and often retire from the same one. This is not the case anymore, especially in the IT industry, in which it is generally accepted that not many employees keep the same job for more than a few years.

The way that architects can achieve job safety is not necessarily by making sure that they keep the same job, but that they make sure they can have a job at any time.

IT architects are in high demand, and if you have the right skills and a good reputation, you will not have to worry much about finding employment. Although this applies to anybody who is good at his or her job, some roles have a better chance of keeping you employed than others. For example, a good mainframe developer can still find a job, but his or her skillset might not be as much demand as that of an IT architect. After all, the architect has the potential to save millions for their company and put it in front of their competitors by choosing the right solution.

How do you prepare for this then? How can you make sure that you are the architect for whom any company is looking? If your current company decides you are replaceable (which can be true), how can you get back on your feet in no time? Here are the things that could make the difference between you and other architects:

1. *The job cannot be easily outsourced (and most of the time is not):* One dreadful word that everybody is scared of: *outsourcing.* These days, the world is much different. As a result of globalization, outsourcing, contracting, downsizing, recession, and even natural disasters, *job security* can sometimes seem like a thing of the past. Nowadays you are competing globally, and salaries for the same position around the globe vary dramatically. Companies may decide to employ a remote worker on another continent if their salary is much cheaper. What can you bring to the table that would shift the balance and will make the company keep you (the more expensive resource) and not hire an outsourced resource? Well, this is one of the times when you have to prove your worth and show why you are the better choice. Most of the time, you prove your worth through the quality of your work and the quality of your solutions. There are a few other things that can shift the balance in your favor including:

   • *Knowledge of the business:* A resource from outside the company has to acquire knowledge of the business in a short time, whereas long-serving employees will already have this knowledge. As an internal resource, your knowledge of the business capabilities, processes, and functions puts you in front as a good candidate for the IT architect role. If one has not been with the company for a long time but has extensive experience working in the same industry, this industry specific knowledge is also a great asset, as the company will spend less money on training. A resource coming from an outsourcing company (assuming that he or she has the industry knowledge) must still absorb and internalize the company's specific business knowledge, as well as the specific technological environment.

   • *Knowledge of the internal culture:* Although this asset is rarely mentioned, knowing the internal culture of an organization is crucial to delivering successful projects. In the case of the architect, he or she will rely heavily on the subject matter experts to produce his or her architecture and to provide appropriate governance to make sure that the solution designed is the one implemented. Simple things like knowing who to contact to get the information you need, and how to approach stakeholders to make sure that they do not feel left out of the solution process, can make a big difference. For example, imagine if you joined a new company and made the mistake of booking a meeting over the lunch hour without explaining why in advance (or providing the lunch). From your perspective, you were just applying the same principles as you did at the previous company where you worked. There, booking meetings over lunch was considered normal. However, at the current company, such actions could keep you from having a productive meeting, and it might take you at least twice as much time to sell your solution. As employees of the company, architects have a tremendous advantage of knowing the culture. They know how they should interact with different people, and they can use this knowledge to their advantage to show how it lowers the overall cost of the project.

2. *Architecture is a job in demand:* This profession has been on the market (in various flavors) for more than 40 years [9], yet it continues to remain in high demand. Why is that? We know that most jobs have ups and downs, and at some point the demand usually drops lower than the supply. However, in the case of architecture, it is not just about acquiring the technical skills and knowledge you need and then starting to apply them. It is also about understanding how to combine those technical skills and knowledge with other skills that will make you successful. In particular, this applies to the inventory of soft skills described in Chapter 5 that include communication skills, organizational skills, and business skills. This unique mixture of hard and soft skills, combined with circumstance that different kinds of architecture enjoy different levels of demand, explains why it can be difficult to find a good architect who would also fit well into the culture of an organization.

3. *The architect as a strategist:* This job involves a lot of strategizing and making sure that the architecture proposed shows business value for each of the work packages that will be implemented. This is especially true for business and enterprise architects. Many companies choose to outsource the grunt work, but not the design, especially when that design is one that confirms that the IT department is fully aligned with the business.

4. *Architecture as a competitive advantage:* Before IT architects were well known in the industry, the value in the architect's position was not really recognized. In 2001 to 2002, high-profile financial scandals at Enron and WorldCom forced the industry to recognize the importance of quality architecture. To protect shareholders and the general public from accounting errors and fraudulent practices, there was an effort to standardize the architecture processes and the way that companies documented their IT architecture [10]. As a result, many companies created their first architecture departments and hired people who called themselves architects (most of the time overnight). At the time, this could occur without the self-proclaimed architects, or the organizations that employed them, having a clear idea of what architecture really was. Today, having competent architects on staff can definitely be seen as a competitive advantage. An architect puts the company ahead of its competitors because the architecture that he or she proposes can make the difference between delivering what the business needs, at the right time, or just delivering what they want whenever they can. Competitive advantage is about using methodologies and frameworks (which we will discuss in detail later in the book) to decide how applications will support the future load, which business processes must be automated, what kind of service level agreements must be implemented, and so forth. In summary, architecture can be considered not only as a cutting-edge profession but also an established one that shows a lot of promising advantages for those who choose to embrace it.

3.4   And (Many) Other Reasons

There are still quite a few other reasons one might wish to become an architect. In case the financial or the job satisfaction aspects did not convince you, here are a few more reasons to enter the architecture ranks:

- *Development path:* As an architect, you will develop your leadership and negotiation skills so that you will be prepared to take on more senior roles in the company. Imagine that you are ready to advance your career, but you have been in a technical position for quite a long time without any chance to demonstrate leadership, or other soft skills, required for such upper-level positions. Interviewers will look for clues that you are able to lead based on your previous experience, and becoming an architect can provide that alternate route to hone your soft skills (leadership, negotiation, and so forth). There are many similarities between architects and other leadership positions such as guiding a team, communicating effectively, and establishing plans.

- *Leadership:* Both architects and other leaders need to have the ability to influence and lead others to achieve valued goals and objectives. A good leader should be able to define a vision and guide individuals towards that vision. In the case of architects, this vision is the proposed solution and the others that they have to influence include a variety of stakeholders such as project team members, account managers, executives, business leaders, and more. Those in leadership positions have to not only create a vision for their team, but must also outline how this vision fits with the company's vision. To be an architect, one should be a good leader as one cannot be successful without being able to influence others, or negotiate a solution.

- *Communication:* The importance of communicating effectively with individuals and groups, both inside and outside of the organization, cannot be emphasized enough. Communication makes the difference between successfully selling the architecture solution and having a great solution that nobody knows about. Strong communication skills make the difference between an effective team that knows the goals and just a group of individuals who have nothing in common other than sharing a common work area and meeting once in a while. In a leadership position, communication is also important because your bosses want you to act as a conduit for information, conveying their message to the staff, checking how they respond, and passing this information back to those bosses.

- *Planning/organizing:* This is an ability to establish and maintain a course of action for oneself and others to accomplish a specific goal by preparing plans and organizing tasks. This is very similar to a project manager (PM) in some ways, as the architect is usually the technical "right hand" for the PM, and they may be needed to provide oversight from a technical point of view on the project. In either case, a leader has to do a lot of planning, which might include preparing the budget, constructing a timeline, establishing and monitoring deliverables, and so on. A leader must also organize the team to make sure that it works efficiently so that each individual plays a role according to his or her strengths. Both PMs and architects have to accommodate all factors in planning work, such as an individual's skills and abilities, timelines, and organizational dependencies in their solutions. Another component of this skill is managing the risks and making sure that one has a clear plan for risk mitigation.

- *Strategic competency:* This is a skill frequently required by executives. Strategic competency enables you to sort through the clutter and find the best route. It is a distinct way of thinking from a special perspective. Although no competency is inherently more valuable than another, this one seems to be a precious gem. Seeing things from this perspective allows you to see patterns where others simply see complexity. You can run alternative scenarios through your mind, ultimately helping you to pick the best solution. Whether this is about the company chief information officer trying to find the best way to use the current resources (either financial or human) to achieve the company's vision, or the enterprise architect trying to come up with the best architecture vision, you require that kind of capacity to rise above the details and to see the patterns.

- *Visibility:* This job puts you in front of the crowd. Usually a company cannot afford to hire more than a few architects, or sometimes you might be the only one in the firm. Although this is true for other roles as well, if you are the type of person who sees a lot of opportunities in the IT architect role, you can easily rise above the crowd. You have access to the important people in the organization (whether it is the account manager to whom you have to sell the architecture for his or her future, or the chief information officer (CIO) who is your main sponsor for the changes for which you are pushing). This brings you one of the most important things that can help you with the financial aspect: visibility. Although the whole team is responsible for the delivery of a certain project (or a program), most of the time you are the *technical interface* with the sponsor of the project or initiative. You will certainly be remembered for a successful project when the year-end comes. As long as you produce an architecture that follows industry-recognized frameworks and methodologies, take into consideration all the needs of the stakeholders, and eliminate the options by enforcing the constraints, this visibility really can be a benefit.

As an architect, you might be one of the most respected and trusted people in the department or the company. This means investing time and effort into building trust with both the members of the project team as well as with the other business and technical stakeholders. The first time that you come up with a solution, they might check and recheck it, and they may still not be totally satisfied. The solution should not be "your" solution; it should be developed in collaboration with the stakeholders. Then, when it comes to selling it, it is just a matter of getting the final nod, and not surprising everyone with something that they have never seen before. If they see their input in your solution, you gain a lot of trust that you can deposit into your "Emotional Intelligence bank" and use for future interactions. As Covey said in his book *The Speed of Trust: The One Thing That Changes Everything* [11],

> …trust is the one thing that can build or destroy every human relationship. The lack of trust will bring down the most profitable companies.

The architect is a counselor, providing consulting services for the executives and business leaders, identifying alternative solutions, and highlighting the risks and implications.

With all these discussions from the first three chapters, I hope I have captured your interest about this marvelous job. In Chapter 4, we will explore the technical background, skills and competencies that can make or break your success as an IT architect.

References

[1] Dale Carnegie Training, "Employee Engagement," http://www.dalecarnegie.ca/employee-engagement/.

[2] Glassdoor, "Business Architect Salaries," https://www.glassdoor.ca/Salaries/business-architect-salary-SRCH_KO0,18.htm, accessed May 2016.

[3] Glassdoor, "Application Architect Salaries," https://www.glassdoor.ca/Salaries/application-architect-salary-SRCH_KO0,21.htm, accessed May 2016.

[4] Glassdoor, "Data Architect Salaries," https://www.glassdoor.ca/Salaries/data-architect-salary-SRCH_KO0,14.htm, accessed May 2016.

[5] Glassdoor, "Enterprise Architect Salaries," https://www.glassdoor.ca/Salaries/enterprise-architect-salary-SRCH_KO0,20.htm, accessed May 2016.

[6] Glassdoor, "System Architect Salaries," https://www.glassdoor.ca/Salaries/system-architect-salary-SRCH_KO0,16.htm, accessed May 2016.

[7] Glassdoor, "Solution Architect Salaries," https://www.glassdoor.ca/Salaries/solution-architect-salary-SRCH_KO0,18.htm, accessed May 2016.

[8] Serra, J., "Salaried Employee vs. Contractor," 2011, http://www.jamesserra.com/archive/2011/11/salaried-employee-vs-contractor/, accessed May 2016.

[9] Perks, C., and T. Beveridge, *Guide to Enterprise IT Architecture*, Springer Professional Computing), reprint of original 1st ed., 2003 ed., New York: Springer, 2013, p. 8.

[10] Wagner, S., and L. Dittmar, "The Unexpected Benefits of Sarbanes-Oxley," *Harvard Business Review*, April 2006, https://hbr.org/2006/04/the-unexpected-benefits-of-sarbanes-oxley, accessed May 2016.

[11] Covey, S. M. R., R. R. Merrill, and S. R. Covey, *The Speed of Trust: The One Thing That Changes Everything*, New York: Free Press, 2006.

4

A Skills and Competencies Profile for an IT Architect

Having gotten an overview of the role and some of its benefits in the previous chapter, it is time now to explore in detail the skills and competencies necessary to become a successful and sought-after IT architect. This chapter provides an overview of how to prepare for a career as an IT architect, and discusses the proficiency levels that differentiate a junior architect (who will not be expected to demonstrate all aspects) from a senior one. As with most other jobs, experience matters. Performing at a high level as an architect involves continually learning new skills and competencies throughout one's career arc. The best architects are always looking for opportunities to build and develop an experience base to increase their value to any organization.

Note that although this chapter details the full range of skills and competencies that are required of a high-performing architect, it is important to remember that it is not necessary to master all of them before embarking on a career in this field. The coverage here provides the reader with all the information that he or she might need to be successful from the beginning. Some of this material may be of use simply as a reference; some of it may be directly applicable to help to enhance specific skills or competencies.

4.1 Technical Knowledge

4.1.1 Common Technical Knowledge

It should come as no surprise that performing the functions of an IT architect requires a significant amount of prerequisite technical knowledge. Most IT architect roles require knowledge of the following concepts at a bare minimum:

- Solid, practical understanding of common commercial information technology concepts, products, technologies, and techniques. Basically, these include all primary concepts and products used in the day to day life of a technology professional, including:
  - Common computer hardware types (mainframe, server, personal computer, laptop, tablet);
  - Common operating systems (UNIX, Microsoft Windows, Linux);
  - General information regarding hardware devices and media for storing and transferring data;
  - Common types of application software (antivirus, office productivity applications, application servers, messaging systems);
  - General networking concepts, including network types (local area network, wide area network, Internet, intranet, extranet);
  - Common network devices (firewalls, switches, routers);
  - General information about computer and network security concepts, including common measures to protect the confidentiality, availability and integrity of the information, creating backups, user types (administrator versus standard), authentication methods, and threat remediation;
  - Software development life cycle (SDLC) concepts, including describing the stages involved in an information system development project, from an initial feasibility study through maintenance of the completed application;
- Knowledge of common enterprise technology domains, including their benefits, drawbacks, and risks. For example, a general knowledge of technologies in the categories in Table 4.1 is an expectation for most IT architects. (Obviously, in-depth knowledge of all of these technologies will come with time and experience, but having a general idea about what they are and how they can be used is prerequisite for any beginning

Table 4.1
Examples of Applications and Systems for Different Technology Domains

Technology Domain	Applications and Systems
Messaging	MQSeries, Sun Message Queue, Tibco
Web and application servers	IBM WebSphere, Oracle Weblogic, iPlanet, JBoss
Business Intelligence (BI) and Analytics	SAP Crystal Enterprise, Tableau, Domo
Customer Relationship Management (CRM)	Salesforce, PipeDrive, Marketo, HubSpot
Business Process Management (BPM)	Total Agility, ECM Documentum, OpenText
Document Management	FileNet, Microsoft Sharepoint
Development, Integration and Modeling	Eclipse, CVS, Subversion, Visio

architect. A basic knowledge of these domains helps an architect understand situations in which certain technologies should be used based on the respective business scenarios.)

- Experience with a variety of business systems. Ideally, a practicing architect will have direct experience working with different business systems such as lending, mortgages, payments, accounting, payroll, and intelligence and analytics systems, among others.

- Broad understanding of the business and industry implications for a wide variety of technologies. An application to access banking accounts for example needs to take into consideration the various business cases such as accessing it through a Web site but also accessing it through Android and IOS apps. This implies using microservices deployed on a light container that can be reused between a Web, mobile app, and even a desktop user interface. The application or solution architect needs to understand each of these business cases and to design the applications and systems to increase the reusability and maintainability. In case of an organization that grew inorganically (through mergers and acquisitions), the architect might want to consider using an enterprise service bus in a federated architecture that would enable the applications across the various lines of business to communicate in a loosely coupled way.

- Ability to understand and articulate the value and technical capabilities of IT systems. Different systems used in the IT industry have varying capabilities. An integration broker system, for example, includes messaging middleware, content-based routing, integration adapters, connection and application protocols, federation services, mapping and transformation, and orchestration, all built in as part of a feature set. An application architect working with integration broker technologies must be able to understand how each of these individual features functions to be able to recommend those which are relevant to the business scenario being considered.

- Working knowledge of the total cost of ownership (TCO) for a wide variety of technologies. Selecting and proposing a solution architecture requires an understanding of not just the best technologies available, but also the ownership cost profiles for those technologies. Total cost of ownership is a critical factor in the comprehensive assessment an architect provides in the selection and referral of information technology (IT) and enterprise technologies. The technology selection should take into consideration not only the hardware and software cost but also the management and support cost, training, and other productivity losses. TCO will be discussed in more depth in a later chapter.

- Understanding of the implications of adopting, changing, or discontinuing the use of technologies. Adopting a particular technology is never an easy decision, especially when that technology replaces or upgrades an existing system or infrastructure. An architect's recommendation to use a certain technology must take into consideration many important factors, such as the credibility and stability of the vendor who supplies it. These are usually defined as part of a reference architecture or reference model. In most cases, this is a prerequisite step in the product selection process. Key to the adoption of a new technology across the enterprise, and thus to an architect, is careful consideration of the costs of migrating to that new technology. For example, if an architect recommends replacing IBM WebSphere as an application server with JBoss, all Java applications deployed on the WebSphere application containers must be modified to support deployment on JBoss and then adequately tested.

If a vendor discontinues a chosen product after a few years, there is another cost associated with evaluating other vendor products and adopting a recommended alternative across the enterprise.

An architect, especially in an independent, client-facing role, is expected to be an expert in the preceding areas. As discussed in the next section, it is important for an architect to possess a breadth of technical knowledge, but also an ability to quickly master specific concepts and technologies when called upon to do so. For example, developers who become IT architects might be experts in the language or application that they use every day, but they will also have to learn how other languages and applications work and interact. In the case of an infrastructure specialist, this might mean that knowing firewalls inside out is great, but that knowing more about the network as a whole and how specific security policies apply to it is more important. It is difficult to "build the house," so to speak, if you don't have the right foundation or technical knowledge to provide a solid base.

4.1.2   Architects as T-Shaped Professionals

Before further discussing the aggregation of the technical knowledge required to be an architect, a brief mention of the "T-shape" personality is warranted. As described in current human resources terminology, a T-shaped person (see Figure 4.1) is someone who combines deep expertise in key areas with broad experience and knowledge across the board. T-shaped professionals are highly valued for their ability to work collaboratively across broad disciplines or teams, while also bringing deep subject knowledge or skills to the table. At one time, such people called generalists, but nowadays are more called versatilists or a generalist specialist.

The T-shape skillset is applicable in nearly all business contexts. Specific to the architect role is the need for breadth of understanding of both technologies and the business systems with which they will be integrated. This is particularly important when articulating the impact of an architecture solution to a cross-functional team. The architect's job is to explain to those stakeholders why a specific solution was chosen, with the goal of receiving full support from all members of the team. Technologists with deep knowledge (depth) in a particular technology or system may not be able to see right away how the proposed technology or system interacts with the others. The level of trust between the architect and the other team members plays a major role in the team dynamic and depends upon the architect's T-shaped skill profile.

Figure 4.1 shows the breadth-versus-depth concept for the various technologies that might exist in the enterprise.

Applying the T-shaped concept to a hypothetical application (or solution) architect in this scenario, it is easy to see that that person has significant depth of knowledge in technologies/products such as Oracle, J2EE, and COBOL but only basic depth in other domains such as networking, storage, and mainframe. In this case, it might be difficult for the architect to communicate effectively with other subject matter experts (SMEs) (either storage or network specialists) to ensure that the proposed solution will fit into the current environment. As an example, assume that the solution outlined above includes a J2EE application, which would make knowledge of the following specialized aspects very important:

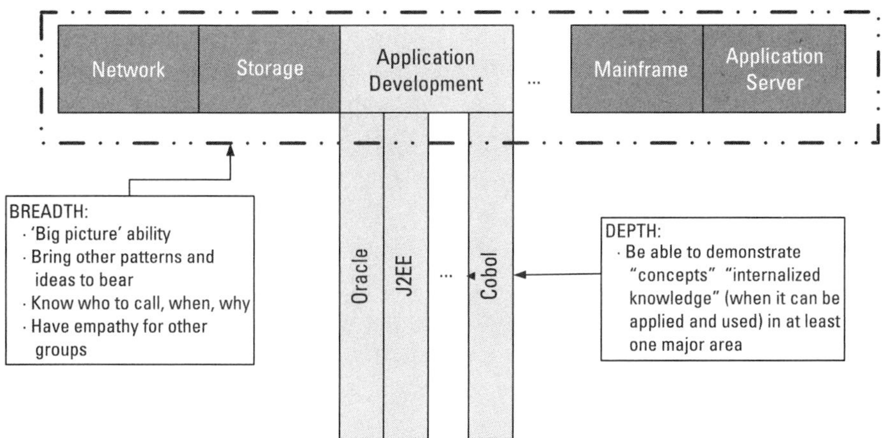

Figure 4.1　T-shaped person.

- Network type: How will the application be accessed? Which protocols will be used and in what network zone should the application be located to ensure the proper security? In this case the application/solution architect proposing the application or system has to have at least a minimum knowledge of the network architecture and infrastructure.

- Storage: Where will the application store the data? Are there any specific storage requirements or any need for replication across multiple storage area networks (SANs)? Are there any solution requirements related to the backup and failover? In this case, the architect should have the minimal knowledge of SAN types, technologies, and failover mechanisms.

- Application server: What is the enterprise technology standard for the application server? What are the details of the deployment process (i.e., should the application be deployed using an automated script that configures all the objects like connection pools and message queues)? In this case, the architect should have some depth related to the respective application server (JBoss, WebSphere, Oracle Weblogic) to understand how to architect the application to take advantage of the specific server chosen for the application.

To reiterate, an aspirant architect should have in-depth knowledge in either one of the application, data or infrastructure architecture domains. Just as having at least one pillar to support the roof of a house is essential, an IT architect needs depth in at least one of the domains to support a solution architecture "roof."

A critical related concept is architect depth. Having depth in one of the domains is important, but it is not necessary, or advisable, to act as the SME for that domain. Architect depth loosely means that, that at most, the role requires only the level of expertise and input required to develop and communicate architecture-level information. For example, an architect who was a software developer in a former job should understand that at *architect depth*, there is no need to get into detail design decisions. Provide breadth and depth, but do not step on the toes of the developer team in the process. A good architect should always include the developer's point of view in a proposed solution; as it is important for that person, normally the SME for that system or application, to buy into its planned implementation. That is because those selfsame developers will be the ones implementing and supporting it. Although this example comes from the application development world, the same concept applies to business, data, and infrastructure domains. Navigating this process can sometimes be difficult. When an architect's specialized technical skills coincide with the application or system on which he or she is working, there can sometimes be a temptation to dive into the details and to take control of the entire depth.

Inevitably, this leads to frustration for other members of the team, who have trouble determining where their expertise begins and the architect's expertise ends.

Another important concept is minimal depth. New architects should be aware that other team members sometimes feel as though a new architect on the project should be the person who "knows everything." Fighting this false perception from the start requires emphasizing that the success of the project depends on all members of the project team and that a broad architecture solution is only as good as the information provided by SMEs. These SMEs are the ones who possess the deep knowledge (depth). The architect's job is to come up with a solution that satisfies both the business considerations and the SMEs' concerns (see Figure 4.2). SMEs possess the depth of knowledge required to document the system inside and out and are usually the best judges of whether a proposed solution will work for that system or application. However, it is important when proposing a truly complementary solution that an architect obtain a minimal depth of knowledge so as to be able to work on the architecture solution for that respective system. Relying on information from various SMEs is absolutely necessary, but it is also critical to achieve a level of depth that allows the entire team to use a common language to discuss the project. This is the best way for an architect to earn the trust and respect of both the business and the SME teams. "Speaking their language" will engender trust in both the architect and the solution that he or she puts together.

So how does an architect obtain this minimal depth of knowledge? Learning as much as possible about the system and the technology before starting on the project is of paramount importance The first impression is the one that

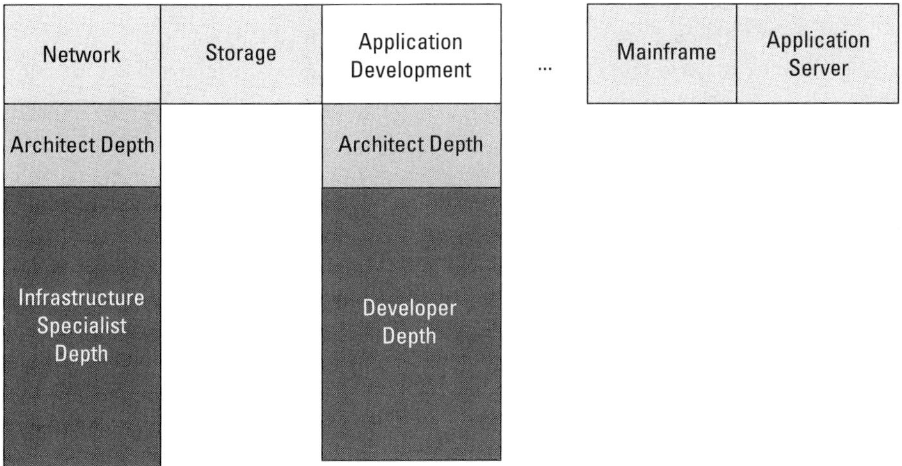

Figure 4.2   Architect depth versus other SMEs' depth.

counts; being prepared for meetings and the first interactions with the project team are crucial. Learning to ask the right questions is also significantly important, and will be discussed in further detail later in this chapter.

4.2 Architecture Design Skills

The ability to effectively design an IT architecture represents the foundation of the skills and competencies that the architect should have. If technical skills represent the foundation for your architect career and soft skills provide the proverbial façade, windows, and roof, the architectural design is the inside of such an edifice.

Architecture design skills may be defined as the ability to create a high-level design that satisfies stakeholder needs and requirements and to present it clearly and capably in documents that can be understood and acted upon by their intended audiences. The scope of this high level design can vary a lot depending on the type of IT architect who creates it. At the enterprise level, for example, such a design will incorporate an entire multidomain enterprise. At the system level, the design will focus on at least one of the systems or domains in that enterprise. An IT architect creates a design in much the same way that a traditional architect would. That is, IT architects create the blueprints that will guide the implementation of the major initiatives for the client company, in much the same way a traditional architect's blueprints describe the construction of a building.

As we can see from Figures 4.3 and 4.4, a high-level design can take multiple forms because it might refer to various different domain architectures. For an infrastructure architecture, the architect would focus on hardware components and certain infrastructure mechanisms such as live partition mobility. A high-level design for solution architecture would focus on different views selected by the architect to address different stakeholder concerns (such as what the main use cases are for the system, what the changes are at the application component level, how the system will be deployed, what are the architectural decisions that the architect had to take during the project, and so on).

Becoming skilled and knowledgeable at high-level architecture design primarily involves first understanding what such a design should look like. The foundation of the architectural design was created with the ISO/IEC/IEEE 42010:2011 standard, which contains the meta-model used by different other architectural frameworks. This standard describes all the elements that compose an architectural model and provides descriptions for associated key terms (model, system, stakeholder, concern, architecture view and viewpoint, and so forth).

The central point in Figure 4.5 is the system, which can fulfill one or more missions. (A mission is defined as the reason why the system exists.) Each

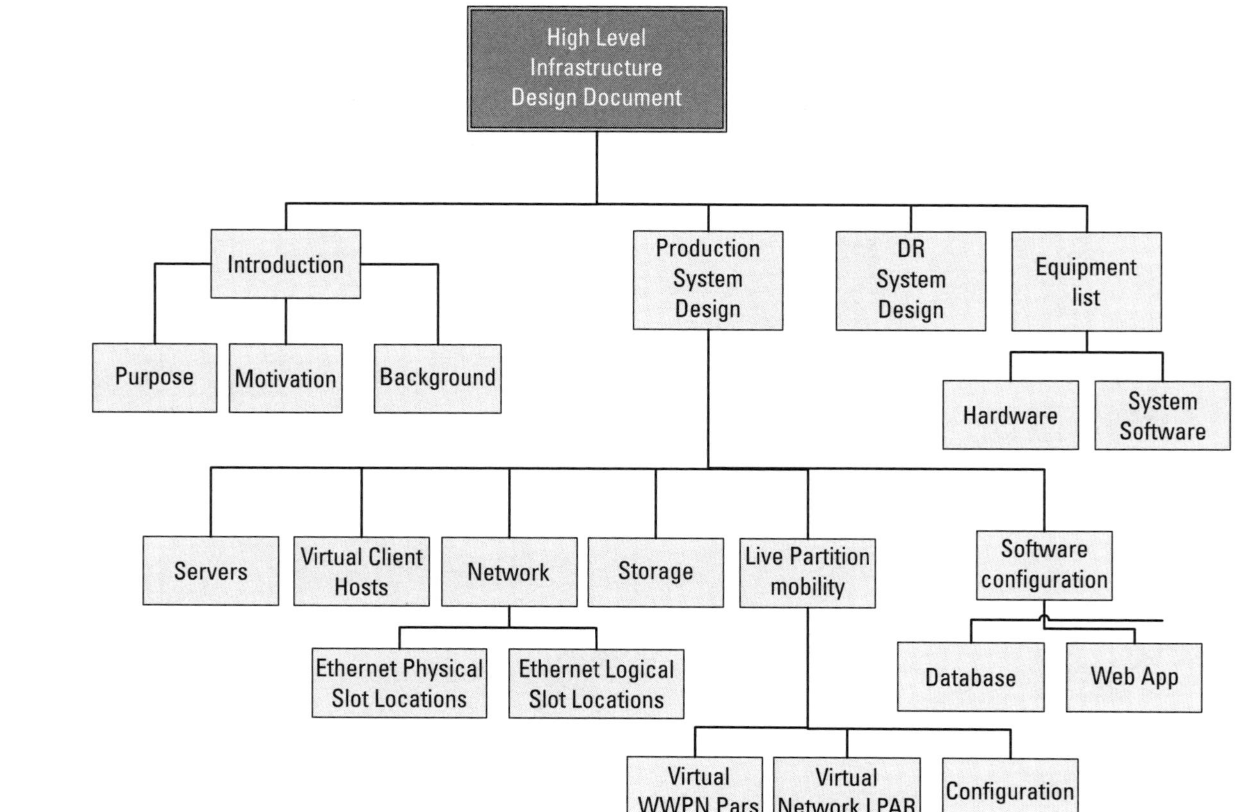

Figure 4.3 High-level infrastructure design document structure.

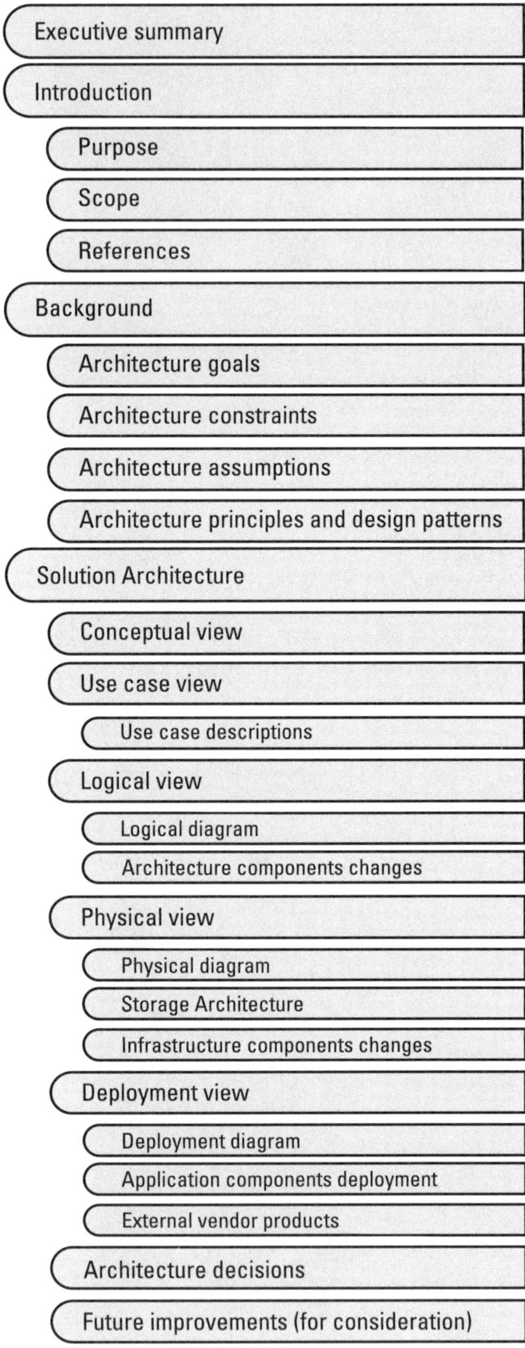

Figure 4.4   Example of a solution architecture document structure.

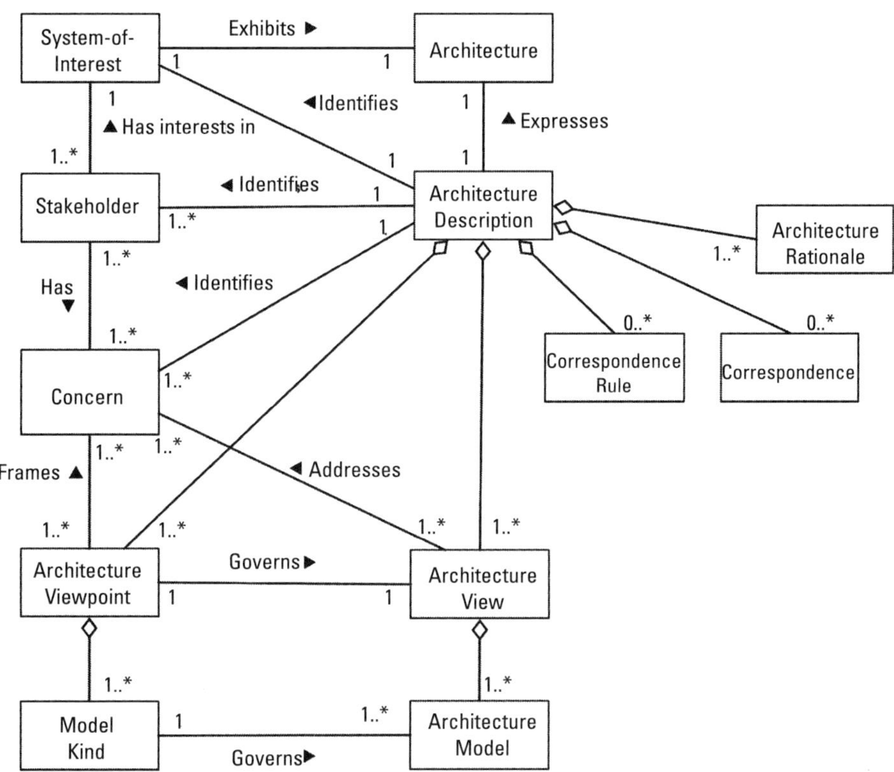

Figure 4.5 Core of architecture description (ISO/IEC/IEEE 42010:2011). (Taken from ISO/IEC/IEEE 42010 website http://www.iso-architecture.org/42010/cm/.)

system has an architecture that is described by an architectural description, which is, in turn, described as a collection of artifacts. The architectural description usually starts with the identification of its stakeholders, who are usually defined as the persons, groups or organizations with an interest (stake) in the system. This includes its clients and sponsors, and the system owner, but also its developers, architect, quality assurance (QA) testers, and analysts, among others. Stakeholders concerns are usually referenced in the architecture design through a set of views and a subset of viewpoints selected by the architect from a library of viewpoints provided in the architecture framework.

Another important concept in this diagram is the model, which contains the modeling objects that are represented in whatever views the architect chooses to provide to his or her stakeholders and other audience members (therefore, that model can be part of one or more views).

We can see a clear separation between the architecture and architectural description as well as the use of viewpoints to create views that become inherent to a model. The focus of this chapter is on the most important architecture design skills, a summary of the elements included in Figure 4.5 will not be discussed here. For a complete description of all the elements mentioned in Figure 4.5, please consult the glossary at the end of this book.

Table 4.2 lists the most important architecture design elements from Figure 4.5:

The modeling elements from the IEEE 1371-2000 form a common base vocabulary (sometime enhanced by additional elements defined by architecture frameworks or modeling languages) that should be used by all the architects to ensure the resulting designs can be clearly understood. These kind of well-defined vocabularies are called architecture modeling languages. Because there are so many kinds of architectures (software, infrastructure, business, enterprise, and so on), there are also a number of modeling languages that span one or many architectural domains. Without getting into a lot of detail (Chapter 6 includes a brief description of the main modeling languages), I will quickly go through the main ones and describe the architectural design skills necessary for using them:

- *IDEF:* A family of languages developed and maintained by Knowledge Based Systems, Inc. (KBSI) used primarily for business and data architecture domains. IDEF's main modeling methods include:
  - Functional modeling: Models the elements controlling the execution of a function and the actors who perform it. It is used primarily to model business architecture and shows the main business functions of an organization and their relations in terms of the information flows.
  - Process modeling: Captures the workflow of the business process via process flow diagrams.
  - Data modeling: Is used to create logical data models. For more information, refer to the IDEF Web site [2], which describes these modeling methods.

Table 4.2
Design Elements ISO/IEC/IEEE 42010:2011IEEE

Element	Description
Concern	A concern is generally considered to be a key interest that is critically important to certain or all stakeholders
View	A view is the representation of a system from the perspective of a set of concerns held by one or more stakeholders.
Viewpoint	The viewpoint is often thought of as the perspective from which an architect looks at the various views and concerns. Most architectures are built from the viewpoint out to address a set of concerns.

- *Business Process Modeling Notation (BPMN):* A graphical language that serves as a basis for business processes and services modeling. BPMN can be quite useful when creating a business architecture that does not need to show any integration with other domain architectures.

- *Testbed:* A business modeling language intended for business processes and organizational modeling. It therefore lacks the architectural perspective of information systems. It is mostly used by business architects.

- *Unified Modeling Language (UML):* An important industry-standard language for documenting various views of software systems. It covers the application and infrastructure architecture and was created initially for the design of object-oriented software. UML is intended for use by system designers, but the diagrams that it produces are sufficiently understandable to be useful and informative for other stakeholders. The main categories for UML diagrams include:

  - *Structure:* Includes the package, class, object, and composite structure diagrams.
  - *Behavior:* Includes use case, state, sequence, timing, communication, activity, and interaction overview diagrams.
  - *Implementation:* Includes component and deployment diagrams.

The beauty of UML is that it can be used for the integration between the application and infrastructure architectures. That is because UML has enough elements to represent the main objects in both architecture domains, which, in turn, enhances traceability. A systems architect, for example, could use UML to represent application services, the application components that realize them (maybe grouped as application functions as well), and the infrastructure nodes or devices where those applications reside.

UML is a great modeling language for systems, application (software), infrastructure, and solution architects, but unfortunately it does not allow for integration with the business architecture domain as it lacks components to represent the main business concepts. The diagram in Figure 4.6 is a classic example of a sequence diagram that identifies the major application components and the interaction among them.

In Figure 4.6 we can see an example of a sequence diagram created to illustrate the flow of the application including any optional branches such as "Client has liabilities with other financial institutions." The sequence diagram is one of the most used UML diagrams. It can be used to show the high-level flow (being a use case realization) or going down to the actual methods/functions that are called (in case we want to show the behavior of the application down to the code level).

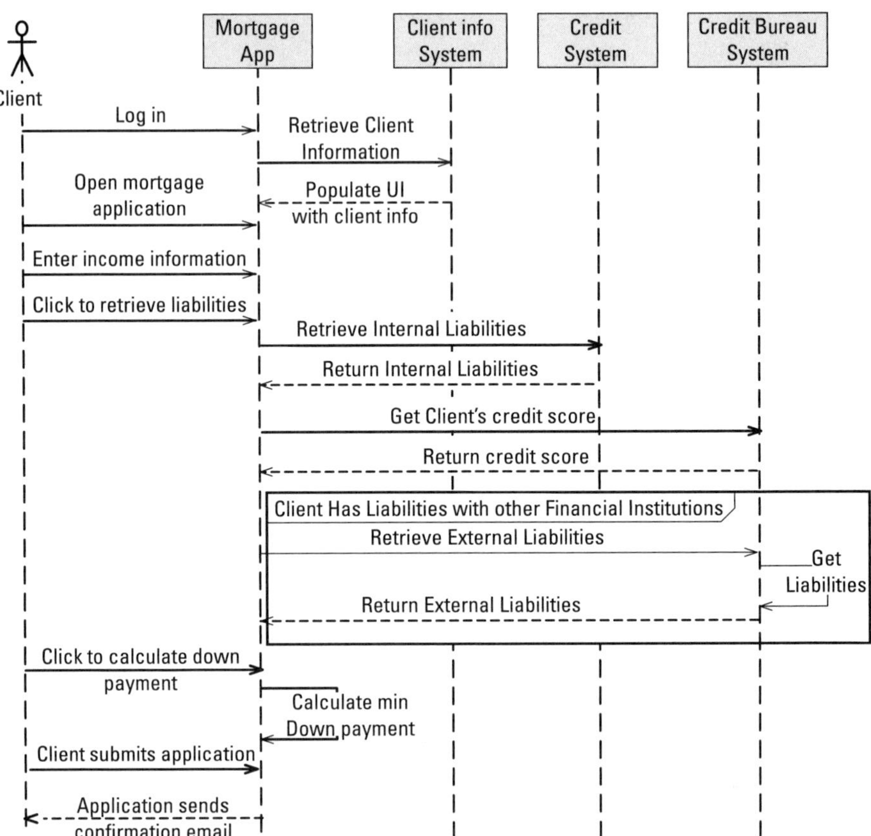

Figure 4.6 UML sequence diagram example.

There are many books available that describe UML in detail, the best of which is *The Unified Modeling Language User Guide, Second Edition* by Grady Booch and James Rumbaugh [3]. There is also a Web site entirely dedicated to the UML specifications, resources, vendors, certifications, articles, and success stories [4].

- *ArchiMate®:* Provides a graphical language for the transformative representation of an enterprise architecture over time (including transformation and migration planning), and also addresses motivation and rationale for changes. The ArchiMate standard is closely related to the TOGAF® architecture framework (for more information, refer to Chapter 7). The language consists of three main types of elements: active structure elements, behavior elements, and passive structure elements.

There are also three main layers that map to the four domain architectures specified in TOGAF, namely [5]:

- Business layer: Documents the business products, capabilities, processes, functions and users.
- Application layer: Supports the business layer with application services realized by application components.
- Data layer: Supports the application layer with database services used by the application components. Although separate layer in TOGAF, the data layer forms one layer together with the application layer.
- Technology layer: Supports the application layer with infrastructure services needed to run applications realized by computer hardware and software.

ArchiMate has the capability to deal with all of the architecture domains in the enterprise architecture, making it the primary choice for enterprise architects who have to document the current and target state for each of the domain architectures as well as the connections between business and IT domains. The architect can focus on those business process that are automated by the various application components through the use of the application services as an interface between the business and the IT (application and technology) domains.

There are other modeling languages in use as well, and an architect you should have a good general idea about each of them to decide which is the best fit for the architecture being designed. In keeping with the earlier breadth-depth discussion, an architect might have depth in a certain modeling language as part of his or her previous experience, but having knowledge of a broad range of modeling languages helps to adapt to the way that other organizations or vendors document their models. As with the human languages, the more languages that an architect can speak, the better. Fortunately, some of the modeling languages might also have common elements (like UML and ArchiMate). In the end, a broad knowledge of the modeling languages will help in designing for general and specific architectures, as shown in Table 4.3.

There are a lot of resources out there with a lot of information about ArchiMate. The classic book is the one published more than 10 years ago by Mark Lankhorst (*Enterprise Architecture at Work: Modelling, Communication and Analysis*). ArchiMate was adopted by The Open Group (which also owns the TOGAF framework), and it is recommended as the preferred modeling language for ensuring an effective enterprise architecture (EA) delivery. There is a dedicated Web site that contains information regarding the certification, pub-

Table 4.3
Design Scenarios and Recommended Modeling Languages

Design Scenario	Modeling Language
Business architecture	BPMN, Testbed, or the business layer in ArchiMate
Software/application architecture (application components)	UML
Data architecture	UML, Entity Relationship Diagram (ERD) language with tool (Erwin)

lications, and events as well as a forum [7]; another forum discussing ArchiMate problems and solutions may also be found on the LinkedIn site.

Regardless of which modeling language is being used, an architect should understand that architecture design consists of describing the architecture (of the enterprise, system, application, and so on) by creating one or more models that encompass multiple views that conform to viewpoints for one or more of the stakeholder's concerns (see Figure 4.7). To do this properly, an architect must be proficient at:

- Setting the scope of the initiative by documenting the assumptions and constraints (technical, budgetary, and so forth) and making sure the assumptions are (ideally) validated before finalizing the design. It is important to understand that this is a critical stage for the architecture assignment. Overlooking constraints or not documenting assumptions might require another architecture iteration.

- Producing a design that:
  - Satisfies all stakeholder concerns.
  - Uses organization specific guidelines, principles, and best practices.
  - Considers both functional and nonfunctional requirements (such as performance, scalability, availability, and reliability). See Chapter 6 for more information on these types of requirements.
  - Minimizes the time and cost of the architecture solution using best practices and proven approaches.
  - Ensures adherence to current architecture standards, strategies, and roadmaps.
  - Provides an architecture design by obtaining required information from various SMEs and then getting their buy-in for the complete solution.
  - Shares the right level of detail and addresses concerns from a wide variety of stakeholders.

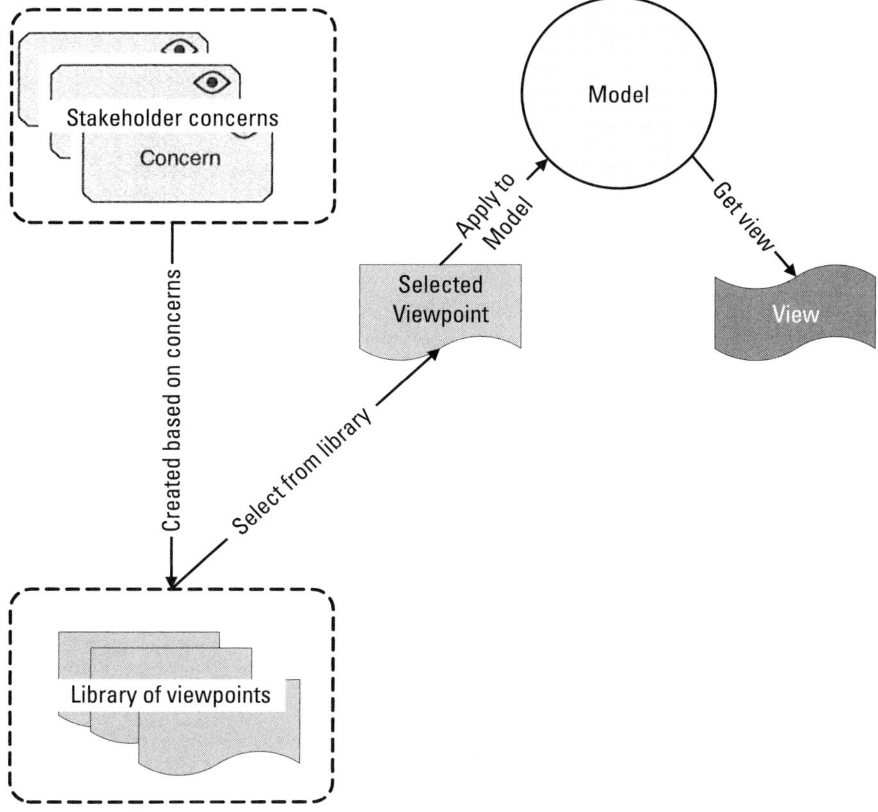

Figure 4.7　Relationship between models, concerns, viewpoints, and views.

- Structures capabilities as granular services, modular architectures, and reusable existing services.
- Uses business knowledge to specify alternatives that better fit the long-term vision, goals, drivers, and constraints of the business.
- Creates and reuses architecture building blocks that improve the time to market for future projects.
- Producing documentation that:
  - Outlines all the architectural information required for the specific architecture assignment.
  - Describes both a system's static structure and dynamic behavior.
  - Produces clear and concise descriptions, avoiding unnecessary verbosity.
  - Is easy to understand for the audience and includes views for all major stakeholders who require an understanding of the architecture.

- Identifies important risks inherent in an architecture and describes them using standard risk-description techniques.
- Defines the gaps between architecture requirements (to get the application or system to the target state) and implemented capabilities.
- Improves the storage and reuse of architectural descriptions so that information is correct, complete, and clear in a way that facilitates making and justifying a decision.
- Follows the architecture governance processes and stores and maintains architectural artifacts and shared models in appropriate repositories. Depending on the architecture level and engagement, the architect might also be involved (working as a consultant) in defining some of the architectural processes, compliance reviews (gates at the major checkpoints in the software development life cycle), coming up with the proposed standards and patterns based on the industry templates, and so on.

The good news is that new architects do not have to "reinvent the wheel" to be successful. The IT architecture practice is quite mature and there are many frameworks and methodologies that can help you produce a good architectural design. We will discuss these in the next chapter, but the important thing here for architecture design skills is that to produce good designs, an architect should have a good breadth covering the architectural frameworks as well to be able to understand which one can be applied in what case. As usual, with breadth should come depth. Most architects should have a solid understanding of at least one architecture framework (Zachman®, TOGAF®, OMG model-driven, and so forth) and know how to tailor it to the needs of a specific architecture assignment. As with the other skills and competencies, following the guidance in this book as well as the other books mentioned in the bibliography will enable you to slowly gain the knowledge and experience required to work on more and more complex assignments that might encompass more than one architectural domain.

4.3   Industry Knowledge

Another important type of knowledge that closely accompanies the technical side is industry knowledge. Technical knowledge enables an architect to come up with technical solutions. Industry knowledge enables customization of that solution to make sure it represents the best fit for the project. Industry knowledge usually consists of concepts and products that are defined outside a specific company. The financial industry is a good example where you have concepts such as bank account, checking, savings, interest rate, and so on that are valid across the entire financial services industry.

There is no commonly accepted definition for industry knowledge. Even so, the following items cover some of its most important aspects:

- Understanding of the industry-related concepts, common products, and product types;
- Knowledge of the major players (companies) in the industry;
- Knowledge of sources of information about the industry including journals, industry-specific resources and process documents, trade organizations, and so forth;
- Knowledge of industry standards, processes, and methodologies;
- Knowledge of the industry's regulatory environment;
- Knowledge of major customer segments;
- Knowledge of the major current trends shaping the industry.

Sometimes a specific industry might have standards, policies, and procedures to which each company within that industry has to adhere. These standards might be imposed through government regulations, or they might come out as a collaboration (or partnership) of multiple companies. In the IT world, there are typically additional levels of standardization, including data models for a specific industry. Some examples include the Pipeline Open Data Standard (PODS), which provides the database architecture pipeline operators use to store critical information and analysis data about their pipeline systems [8] and ARTS data model, which provide retailers and vendors with a mature data architecture for developing retail business solutions [9].

Having this industry knowledge is really beneficial to architects, who usually have hands-on experience with the major pain points in the industry and most likely were involved in crafting solutions to such problems. Architects should understand the dynamics of the industry, how money is made, what the costs are, what the major trends are, what the points of business friction are, what partnerships need to be secured, who the key players are, how to network, and how to build credibility. Researching and understanding trends in technology and in other industries to see how to innovate on those problems in a way that creates significantly more value than the cost of implementing any business change are also important skills to master. (This is especially important for enterprise architects, who often must put together enterprise solutions that satisfy business goals, vision, and strategy that are most often benchmarked against industry metrics.)

Using the banking industry as an example, let's look at some specific knowledge the architect should have to be able to produce successful technology designs that solves the business problems:

- Common banking products such as mortgages, loans, and financial planning, as well as most common concepts used such as customer and account information and major business processes (account creation, line of credit application, and so forth).

- Banking industry regulations might differ from one country to another. As an example, Basel III specifies rules for capital (Capital Adequacy Requirements Guideline) and for liquidity (Liquidity Coverage Ratio with its 30-day horizon, and the Net Stable Funding Ratio with a time horizon of 1 year).

- National regulations: In the United States, banks are regulated by and must be members of the Federal Reserve System; in Canada, the regulatory body for the banking industry is the Office of the Superintendent of Financial Institutions (OSFI).

- Journals and publications for bankers such as ABA Banking Journal, American Banker, Bank Director, Banking Strategies, Banking Technology (information source for systems purchasing in the banking industry), and Global Finance.

- Common banking processes and procedures such procedures for controlling cash, opening checking accounts, check reconciliation in between the banks, assessment and adjudication of a mortgage or lending request, and so forth.

Industry knowledge can generally be classified into two dimensions, namely, *time* (current versus future trends) and the *breadth* (concepts, products, standards that might apply to the various industries), as we can see in Figure 4.8.

As with technical knowledge, industry knowledge (and business knowledge) has a breadth dimension that tends to be overlooked most of the time. People work for most of their lives in the same industry (or similar ones), so they tend to have a lot of depth into the industry knowledge for that industry. There is very little knowledge across different industries; even an enterprise architect only looks at the enterprise within the context of the client company's particular industry. Most client or hiring companies often give preferential treatment to candidates who have worked in the related industry for at least a few years. As an architect, demonstrating breadth of knowledge across industries is very important. Sometimes, the problems that need to be solved in one

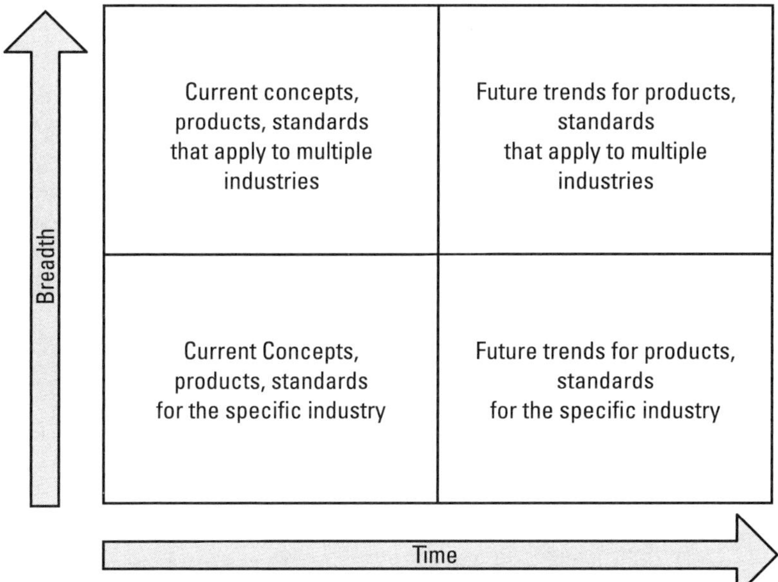

Figure 4.8　Industry knowledge dimensions.

particular industry already have a solution in a different one. Borrowing ideas from other industries and disciplines is one of the best ways to make breakthrough improvements. Coming up with an effective solution company (application or system) requires the ability to look outside for ideas and best practices or other ways to achieve maximum performance in any given area. An architect should be able to broadly define a problem, search out the various ways others have solved that broad problem, and look at the solutions that already exist as either a direct or partial solution. For example, Intel is currently looking to borrow technologies from the aerospace industries (specifically aircraft design) to develop better casing for laptops by laying out their internal components in ways that strengthen the structure. This seems to offer a partial solution to a problem that, when combined with other innovative design elements (such as ways to dissipate the heat faster), will eventually lead to sturdier laptop cases.

4.4　Business Knowledge

Business knowledge is defined as a thorough understanding of the general business capabilities, functions, processes, and services of an organization. There is a tendency to consider business knowledge a subset of industry knowledge; however, there are certain aspects outside of industry knowledge (such as the capabilities and functions specific to the company).

There are concepts, product types, and terms that are specific to the industry (please refer to Figure 4.9). Then the business knowledge specific to the company represents a refined and more concrete form of knowledge that is specific to the company alone. The business capabilities and functions might be common across an industry (handling claims or asset management in the insurance industry, for example), but there might also be others that are company-specific (they might be also related to the way that the company is organized). The accuracy and completeness of this knowledge are crucial to the development of a strong and solid foundation upon which an architect can build an understanding of the user and the user's problems and design new and more efficient methods for solving them.

As it relates to an architect's design skills, the following question might well be asked: "If this kind of knowledge is so crucial, how can it be obtained and used it in conjunction with technical skills to produce the best architectural solutions?" The unfortunate answer is that most of the time there are no such documents, Web sites, or other sources that can give a new person in an organization a shorter ramp-up time. Yet this is essential knowledge for an enterprise architect, for example, to use come up with an overall vision. It is equally important for an application architect to use this information when crafting a solution that is a good fit for the respective business capability. This knowledge is often accumulated over many years of working in the same company and the same business unit, but it can also be obtained through coaching by senior peers or by managers or indirectly from the requirements or business needs documents that are produced (hopefully) during other projects. In this case, possessing industry knowledge would certainly help because you only have to map the industry products/concepts/business functions to their corresponding elements within the company.

As an example, a company from the banking industry would have different lines of business (LOBs) such as loans and mortgages, payments, credit cards, and deposits. For each of these LOBs there are different business concepts that an architect would have to know, to wit:

- What are the main capabilities, products and services offered by each LOB? There are different account types such as checking and savings, each of which is associated with different services (overdraft protection and insurance), credit products (credit cards, lines of credit, personal loans), types of payments and the interest associated with them, and different check types (traveler's checks, money orders, and bank drafts).

- What are the bank-specific processes and procedures associated with these products? Such processes include how to open a client account, how the client information is stored and accessed, what the processes are to adjudicate a loan or mortgage, and so forth. Each of these will map

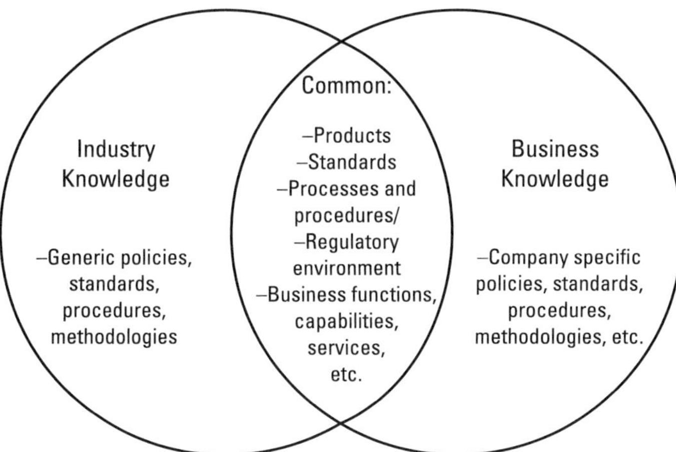

Figure 4.9   Industry versus business knowledge.

to the business systems and the architect has to know them well to be able to apply certain principles to provide the best solutions. As an example, one of the business principles might be ensuring a flawless client experience and, as a result, the architect would recommend storing the client information in a central place so the client will not be required to provide the same information multiple times.

• What are the other financial products outside of banking offered by the bank (mutual funds, investment accounts, insurance products, trusts, private wealth management)? What are the common business concepts shared with the bank (see the customer account example earlier)? This will allow the architect to understand the full business picture and be able to recommend solutions that will reduce the cost by increasing the reusability of the IT resources (such as application services).

The architect would not be able to provide a solution for the various domain architectures without having at least a general understanding of these concepts. He or she will not delve into more details unless this would be required by the system or application supporting the business functionality.

Depending on the level of architecture designed, an architect might also need to know something about the major competitors and partners because the business knowledge coming from this direction might help with finding new ways to improve business operations. Similar with borrowing industry knowledge from other industries, there is the concept of borrowing business knowledge from other companies. This is not so much borrowing as it is trying to find the business knowledge concepts that made another business successful and using them to define new business architecture.

References

[1]   ISO/IEC/IEEE 42010 Web site, http://www.iso-architecture.org/42010/cm/, accessed May 2016.

[2]   IDEF web site, accessed May 2016, http://www.idef.com/, accessed May 2016.

[3]   Booch, G., and J. Rumbaugh, *The Unified Modeling Language User Guide*, 2nd ed., Boston, MA, 2005.

[4]   Unified Modeling Language Web site, http://www.uml.org/, accessed May 2016.

[5]   TOGAF Domain Architectures, http://pubs.opengroup.org/architecture/togaf9-doc/arch/chap02.html, accessed May 2016.

[6]   Lankhorst, M., *Enterprise Architecture at Work: Modelling, Communication and Analysis*, Berlin Heidelberg: Springer-Verlag, 2005.

[7]   Archimate Forum, http://www.opengroup.org/subjectareas/enterprise/archimate, accessed May 2016.

[8]   PODS Web site, http://www.pods.org/, accessed May 2016.

[9]   ARTS Data Models, https://nrf.com/resources/retail-technology-standards/data-models, accessed May 2016.

5

The "Soft" Background of an IT Architect

So far, this book has focused on the knowledge-based skills that an information technology (IT) architect needs. They represent the core of any architect's profile because they are necessary simply to understand the *IT language* that is spoken in the trade. For example, there is an enormous gap between the level of knowledge and understanding that is needed to understand the jokes that native speakers may make. As with any other language, knowing how to speak and understanding what others are saying is only the beginning. Advanced understanding and nuance is needed to grasp the historical, geographical, and political context of humor.

The same is true for the architect's role: here, too, possessing technical skills and industry and business knowledge is only a beginning. To be a successful architect, many more skills are needed. Most technical problems will require a consistent approach to solve or resolve them. An architect needs to know how to communicate his or her solutions, how to adapt his or her communication style and material to the audience, and how to influence others to achieve his or her objectives.

In this book, such skills are all grouped under the *soft* background of an architect (also called *performance competencies*). That is because they constitute the set of behaviors or actions that need to be performed successfully within a particular context to achieve the best outcomes. In the following sections, we will review those skills in some detail.

5.1   Analysis and Problem-Solving

Because an architect has to select any solution based on rigorous problem solving, analytic and problem-solving skills sit at the borderline between core competencies and soft skills. These include an ability to gather relevant information, to utilize various problem-solving tools and methods, to elicit accurate implications, and to demonstrate creativity and innovation when reaching decisions or solving problems. Needless to say, the definition of the problem may vary at different levels of the architecture: for enterprise architecture, the problem might be how to achieve proper integration between headquarters and business units to achieve consistency and implement uniform processes and procedures at each site. For infrastructure architecture, the problem might be how to zone the network to ensure a balance between proper security and ease of access. Each architecture has its own type of problems, but an architect should be able to use a structured approach to analysis, problem-solving, and decision-making for any and all of them.

To be able to solve such problems, an architect must be able to employ certain techniques that can help him or her arrive at the best solution. These techniques include:

- Collecting information from all relevant sources and asking effective questions to elicit this information. Gathering information is often a critical part of analysis because without the right information, the architect may not be able to pinpoint the root cause of the problem and to select the right option to address that root cause correctly.

- Problems may also be classified by their complexity, by how well defined they are, and by their scope. It is always important to decide how much information is enough: sometimes, we tend to collect more information than we need for the problem we are trying to solve. Beyond a certain level, although confidence increases, accuracy actually decreases. One good way to avoid overkill is to document one's findings and continually question all prevailing assumptions.

- Select a problem solving method based on these different classes of problems:
  - Creative: brainstorming: This is great for simple problems and as a way to increase the trust and the team spirit. That is because brainstorming makes project team members feel included and more inclined to take responsibility for the final solution. Brainstorming is a popular tool that helps to generate creative solutions to a problem. It is particularly useful when there is a need to break out of stale, established patterns of thinking to develop new ways of looking at things. It also helps overcome many of the issues that can make group problem-solving a sterile and unsatisfactory process.

- Rational: The Kepner-Tregoe problem solving technique is good for structured problem-solving with complex, but well-defined problems [1]. It is not the best approach when the scope is not clear or there are other human factors involved.
- Soft systems: Soft Systems Methodology (SSM) is a better approach for unstructured problems that require an understanding of the views of the key stakeholders [2]. For this particular type of problem, the iteration between the internal thinking and the real world is essential to validate the conceptual models created. One of the interesting things about SSM is that it constrains thinking so as to expand thinking, thus debunking the notion that systems thinking is always expansive.

These are just a few examples of some common problem-solving methods. There are many other methods that may be used in different circumstances. Because problem-solving is such a key skill for any architect, it is critical to become familiar with these methods and others like it.

Once a problem-solving method has been selected, it can be used for any kind of decision-making including the *architecture decision document* deliverable discussed in Section 7.2.2. Decision-making is based on problem-solving skills. Thus, any decision made that relates to a final solution should adhere to some common, well-known method (such as one of the methods described earlier in this section). Problem-solving is an important skill, but one should not make decisions without applying proper architecture design skills either.

One of the important steps in the decision-making process is generating alternatives using lateral thinking and producing ideas without an initial thorough analysis to see which one provides the best fit. This helps when coming up with unusual or different solutions that might prove more successful than standard ones. As an example, when trying to reach a decision about where to put an application, it makes sense to first use divergent thinking to try to see what other patterns are out there and what has been used for similar applications. Research might lead to the discovery that virtual servers can be used on an existing physical machine. Thus, an existing solution building block can be recast, rather than going through the hassle of buying, configuring, and maintaining a separate machine.

Once enough alternatives have been collected (remember to always question how many are needed), the evaluation process can start. To provide a fair evaluation, a weighted matrix that contains the criteria and their associated weights will usually be used. The important thing is to pick criteria that are relevant to the problem. Such an approach will differentiate among alternatives and allow the best one to make itself known. Table 5.1 provides one example for some primary criteria including scope, compliance, budget, and nonfunctional

Table 5.1
Example of Criteria for an Architectural Decision

	Option 1: Microservices Deployed in the Docker Container	Option 2: SOAP Web Service Deployed in JBoss Fuse
Business strategies		
Business risks		
Compliance		
Budget		
Scope		
Quality		
Schedule	This is the first implementation of microservices in the organization; best practices and patterns need to be developed.	Schedule will not be impacted as the organization has already best practices and patterns for SOAP implementations in place.
Delivery risk		
Technical risk		
IT operational cost		
Business operational cost		
Technical risk		
Training	Requires training as the development team does not have the necessary knowledge	No extra training is required as the development team has extensive experience JBoss Fuse.
Solution direction	This option is in line with the mid-tier strategy closing the gap to achieve the target state.	This option maintains the status quo and does not close the gap to achieve the target state
Organizational compatibility		
Availability		
Maintainability	Easier to maintain as a defect or a functionality change affects a limited number of services.	More difficult to maintain as all the services are incorporated in one EAR file which has to be redeployed every time a change is implemented
Manageability		
Performance	Microservices are light services that can easily handle an increase in workload.	More work is required to properly adjust the performance of the SOAP services.
Reliability		
Recoverability		
Component		
Disaster		
Accessibility		
Adaptability		
Interoperability		

Table 5.1   (continued)

	Option 1: Microservices Deployed in the Docker Container	**Option 2: SOAP Web Service Deployed in JBoss Fuse**
Scalability	Microservices scale very well, especially when deployed in a virtualized/cloud environment. New instances can be started in seconds	Not so scalable and flexible, formal verbose contracts, state full with bigger payload, does not scale so well.
Portability		
Extensibility	The functionality can be easily extended as each piece of functionality is encapsulated in a separate service.	Extending the SOAP service and adding more methods requires redeployment of the whole ear file and regression testing.
Assurance		
Auditability		
Security	Requires more work to secure the services including the communication between them.	SOAP has protocol extensions WS* to ensure the security of the information transmitted.
Privacy		
Integrity		
Credibility		
Usability		

properties. Do not add more criteria unless they will produce a difference between options, or unless they are especially relevant to a decision. Table 5.1 illustrates example criteria for an architectural decision.

Once the options are listed, it becomes easier to eliminate those items that are not really mandatory criteria. Then a more detailed analysis can be undertaken to come up with pros and cons for each remaining option and, finally, to use that information to select a solution. Stakeholders and team members should be involved in selecting criteria and in defining their weights. This will help in *selling* a resulting solution to them once a decision is made. Sometimes, that decision may need to be revisited because certain project constraints might change, some assumptions might be invalidated, or because new facts come to light.

5.2   Inquisitiveness

Inquisitiveness can be considered part of the information collection phase for problem-solving. However, there is more to inquisitiveness than just the way we collect the information to solve a particular problem. This can be defined as [3]:

> ... given to inquiry, research, or asking questions; eager for knowledge; intellectually curious: an inquisitive mind.

The most important part in the definition above is "eager for knowledge," and this is an aptitude that all architects should possess. The best architects are those who know which questions to ask and how best to ask them.

What about other skills? Simply put, they are equally important, but without the inclination to be inquisitive, it might not be possible to get the right information at the right time. The right information is that which makes us choose the right solution and the right time is about getting this information during the design stage and not later on during implementation.

How is it possible to train to be inquisitive? Although numerous training courses focus on facilitation, inquisitiveness is not just about facilitation; it is also about assessing and documenting current and proposed architectures. Converting information into useable knowledge requires a repeatable, structured approach to gathering information from internal stakeholders and documents, as well as focusing research on publicly available product and industry information. Inquisitiveness is also mostly about the technique used to collect the information both internally and externally. This information needs to be validated with the stakeholders by first documenting it in a set of diagrams that should represent a specific view (according to the selected viewpoint). The whole process is an iterative one because most of the time the information collected in the first iteration makes people think about the processes, components, and services and provides them with a structured model that is much easier to understand and validate.

As an architect, you need to be able to draw. This is the part that is in common with traditional architecture, although the drawing required from an IT architect needs to only present concepts in a visual form. There is a well-known saying that a picture is worth a thousand words. This has been confirmed by recent studies that show that the human brain remembers pictures better than words [4] and that each word is recognized as an individual picture [5]. Drawing an architecture solution instead of explaining it in a few pages of words is the preferred approach, at least for initial iterations when there is no agreement on the design. After a few iterations, once everybody is in agreement in regards to the current and target state of the architecture, then the architect can certainly add more text to document a solution. Most of the time, the architect would draw diagrams even when collecting the information because this would help with an early validation from the person providing that information.

Inquisitiveness has another interesting component: getting into the habit of understanding the why behind the solution instead of mechanically coming up with a solution based entirely on patterns and frameworks. An architect is expected to be able to ask the right questions to get the information needed to pick a solution, including:

- Why do we need to automate some of the business processes and leave some others manual?

- Why do we even need the project? Is there another way to arrive at the same outcome without spending time and money?

- Why do we need to implement the functionality in this system and not in another one? This is often a point of contention between development teams.

- Why do we need to think about nonfunctional properties of the system? Will the system really have performance or reliability problems?

These are just a few examples, but most experienced architects would be able to come up with many more just like them. An architect has to preemptively find out the whys specific to the project or initiative so they can be incorporated into the solution, but also so they can answer such questions during reviews of the architecture document or later on during implementation.

Finally, this is one of the skills that it is acquired with experience. There is no manual on which questions should be asked for each scenario and this is the reason why this is a skill and not a competency.

5.3 Leadership

None of the definitions for leadership that I found were a good fit for the IT architect role. In my opinion, one of the essential components of leadership is the ability to influence and lead others to achieve common goals and objectives [6]. This is done mainly by communicating information or arguments in a manner that gains agreement or acceptance and by being able to facilitate discussions to solve potential points of conflict or contention. Leadership is the main skill in common for the architect and for management as both should influence people towards some common good. In the case of the architect, this is actually the architecture solution that will be implemented, whereas for the manager it is about leading the team. Although we use the same name, the leadership skill that the architect needs to have is quite different from the skill that the managers have. The architect needs to lead mostly by influence without being in charge of the team. He or she has to employ quite a few techniques and exhibit a few personality traits that will increase trust in him or her and make others follow his or her directions for the implementation. The important thing to understand from the beginning is that the architect has to be a strong leader to be successful in influencing other members of the team.

The choice of IT architecture can be a political issue, and so leadership and consulting skills are important. Leadership is not only about managing personnel to achieve technical value but also about stakeholder management. The architect is supposed to:

- Communicate, promote, and demonstrate the value of his or her vision or solution to appropriate levels within the organization.
- Build relationships that assist in being an effective leader, resolving difficult issues, and furthering important initiatives.

The following elements are the main components of leadership from an architect's point of view:

- First and most important, *trust:* A leader needs to inspire trust and this means that every member of the project team needs to trust the architect. This is easier said than done considering that sometimes the architect is either new to the team or even new to the company. Being either a new employee or a contractor does not help much with trust as people may assume that the architect wants to impose his or her solution without consulting them. An architect should never impose leadership ("I am the architect here, so you should implement this solution") but get things done through personal influence and credibility. Trust is built in small steps and can be lost very quickly. Without trust, there is no way to move forward. How can an architect become trusted by others and become a great leader? Here is a summary of some of the most important character traits for building and maintaining trust:
  - The most efficient way to build trust is to have a consistent track record of delivering results and doing things as promised. This is very much connected to integrity, which has become so rare today in a world where everybody is focused on results, sometimes even compromising their integrity or "bending the truth."
  - Show respect. This is a great way to gain trust; we are social living beings, and as a result, one of the most important needs (apart from biological ones) is the need for respect. Showing respect to the subject matter experts for various systems satisfies that need and therefore increases their trust.
  - Be transparent. It is not always possible to be totally transparent but be as transparent as possible because even the suspicion of withholding information might put a serious dent in a team's trust. Be explicit about what information is not allowed to be shared.

- Extend trust. To be trusted, you must trust others yourself as well. Trust is earned but at some point one person has to make the first move and trust the other, so be the first.
- Keep commitments. State your commitments and keep them in front of everybody.
- Confront reality. Great leaders do not avoid conflict; they tackle it head on. As an architect, there will be conflicts that have to be addressed. Other members of the team may disagree with the solution proposed for various reasons. Avoiding the conflict and maybe (even worse) getting the support from the project sponsor to impose your solution is the worst thing to do. The only way to move forward is to get into the conflict, explain the benefits of the solution, and acknowledge the drawbacks. Getting back to inquisitiveness, the why is one of the basic questions and it has to be satisfied for the team to buy into the solution and achieve a successful implementation. Getting the sponsor to push for a solution from the outside of the team will only temporarily solve the disagreement and, in the long run, will prove worse as the conflict remains unsolved.
- Demonstrate that you are one of the team. Although you might have a totally different background, showing others that you know what they are talking about is a great way to increase trust. Before joining the team, read about the system, the application, the infrastructure, the business processes, or whatever you might need to know for this assignment.
- Increasing trust means demonstrating all of these behaviors to both project team members as well as project sponsors (account managers and chief information officers).

- Clarify expectations and goals: Paint a compelling picture that everybody will keep in mind at all times.

   This leadership component means building on top of the trust that you have achieved and showing the team the final goal. There is a huge difference between a team that has a clear technical goal for their project and one that does not. The architect is like the general walking in front of his troops and telling them exactly why they are fighting the battle. The better the speech, the better the chances to win the war. The architect is the technical lead for the project and should not only know exactly what technical value the project is bringing to the organization but also what the successful outcome is. He or she should be able to communicate this to the rest of the team and the sponsor (communication will be covered later in Section 5.4).

- Be a role model: Showing respect is one component of trust, but getting respect from with a team can only be done by being a person whom they would like to become. Leadership is also about knowing *your personal style* and trying to find out as much as possible about others' styles. Personality traits are beyond the scope of this book, but there are quite a few personality profiles that help to determine different personality traits. Understanding these different profiles improves interaction with others. One of the most popular is the DISC personality profile [7], a personality assessment tool based on the theory of psychologist William Marston that centers around four different personality traits: Dominance, Inducement, Submission, and Compliance. One field in which DISC assessment can be used is leadership because there are different leadership methods that could help leaders be more effective. DISC has also been used to help determine a course of action when dealing with problems with team leadership and individual roles—that is, taking the various aspects of each type into account when solving problems or assigning jobs.

As a leader, it can be helpful to determine the personality type for each team member and then to approach the team member according to his or her desires and tendencies. For example a "D" type of person needs to have their power and authority acknowledged, needs direct answers and freedom from controls and supervision as opposed to an "S" type of person who loves to maintain the status quo, have predictable routines, and standard operating procedures. An understanding of the different motivational environments required by the various behavioral types can help a leader provide all the information required for each of these types.

No matter the level of technical skill or years of experience that an architect possesses, improving leadership skills together with the other soft skills covered in this chapter will definitely improve his or her chances of success. As with any other performance competencies, leadership skills can be improved by taking leadership courses geared towards the specific needs of each type of architect. The American Management Association (AMA), for example, provides different leadership courses to help attendees to become better leaders and to better equip them to assume the responsibilities of new roles. Preparatory courses can also explore facets such as leadership strategies (which are required by enterprise architects), communications (to be able to clearly communicate and get the buy-in for solutions), or the coaching or mentoring of team members (to be able to lead the development team, for example, as an application architect to ensure the implementation according to the planned architecture).

The AMA, along with numerous other global course providers, also offers numerous courses that would be helpful to those preparing for an IT architect role. Such offerings include how to hone critical thinking and how to develop collaborative leadership skills to inspire the best performance from team members and work with them to achieve the best solution architecture through collaboration. The AMA, for example, has courses specifically aimed at technical professionals that could help anyone interested in becoming a successful IT architect (Leadership Skills and Team Development for Technical Professionals). For more information, please visit their Web site: http://www.amanet.org/training/seminars/leadership-training.aspx.

To summarize, leadership is by far one of the most important skills for an architect as he or she is expected to lead the project team from a technical point of view. That it is only one of the pieces of the puzzle, as there are quite a few other skills and competencies required. Although the information discussed in this section might seem overwhelming, the important thing to consider is that leadership is a skill acquired over many years and a junior architect could not be expected to demonstrate all aspects. It would be beneficial to be aware of the skills required though and to work continuously to keep developing and improving them.

5.4 Communication

Communication is defined as the activity of conveying information through the exchange of thoughts and messages by speech, visuals, signals, writing, or behavior [8]. Together with leadership (and sometimes considered part of it), communication is one of the most important skills for an architect. As we saw in previous chapters, technical background is important to make technical decisions, but without proper communication it will not be possible to get the information necessary for the technical solution and to sell it to the stakeholders. Communication skills (or the lack of communication skills) can have a large impact on success. Even in other IT roles (as a developer, for example) where there is a bigger emphasis on the technical competencies, communication often makes the difference between successful and failure.

The following types of communication have special meaning for IT architects:

- *Oral communication (spoken or verbal communication):* This used to be face-to-face communication that would give you immediate feedback. Nowadays, this can also be a telephone conversation in which it is important to be aware of other cues, including tone of voice. Verbal communication is very important for an architect to collect the information that he or she needs to crystallize the architecture solution but also to deliver successful presentations once the solution is ready.

- *Written communication:* Allows you to communicate messages with clarity and ease to a far larger audience than through face-to-face or telephone conversations. This is crucially important for architects because they have to deliver an architecture solution in written form. Poor writing skills create poor first impressions and many readers will have an immediate negative reaction if they spot a spelling or grammatical mistake. The focus in this case might shift from conveying technical information to making sure the language of the message is correct. Sometimes a reader might also make assumptions that the content is only as good as the delivery container and poor written communication might mean an architect has to spend a lot of time trying to rectify this assumption.

  Under written communication, we also include presentations that an architect is likely to need to produce and deliver. These are some of the most important ways to deliver information, and as a result, it is important to be careful about the kind of information included as well as the delivery format. Consider the audience before making the presentation itself and where the audience is unknown at least try to simplify the vocabulary so that the majority of the audience can understand it. In a later chapter, presentation skills will be discussed in more detail because they are also crucial to success.

- *Nonverbal communication:* This does not mean written communication. It is also known as body language. All of our nonverbal behaviors—the gestures we make, the way we sit, how fast or how loud we talk, how close we stand, how much eye contact we make—send strong messages. Nonverbal communication is constant and applies to speakers and nonspeakers. There are many types of nonverbal communication, including facial expressions, body movements, gestures, eye contact, and voice. For the architect, nonverbal communication is important as people will watch him or her during presentations or other meetings to try to gain more information about his or her personality. A mismatch between the verbal message and the nonverbal ones might be also an important clue that something is fishy, and as a result, others might lose their trust in the architect.

  Nonverbal communication takes precedence most of the time over verbal communication as it is basic human nature to find behavioral cues and to try to guess what is behind the words. For example, when trying to collect more information for the solution architecture, if you assure the person with whom you are talking that you are very interested in what he or she has to say but you yawn, fiddle with your fingers, or (even worse) look somewhere else, the other person would definitely think that you are not being completely honest. It is not possible to totally control our behavior and it is inadvisable to try to mask all signals like this, but be aware of the effects that this (sometimes involuntary) nonverbal communication can have on the perceptions of others and adapt accordingly.

Previously, we ran through various types of communication and how they are useful to the architect in the day-to-day life. Next, let us take a closer look at how we can develop effective communication skills:

- *Listening:* This is one of the best ways to get the most out of a conversation and also increase trust between those in a conversation. Listening means not only hearing and understanding the words being spoken but also understanding how they are being spoken and the nonverbal messages sent with them. What you hear may not be what the speaker actually meant, so most of the times, paraphrasing and restating both the feelings and words are a good way to ensure that both speaker and listener are in agreement. This is an important component of active listening that also includes making eye contact (and maintaining it) and reacting (e.g., nodding) to what others are saying. This is important, as everybody looks for cues that they are understood as they are speaking. If listeners do not give any indication that they actually understand what is being said (or if they agree or disagree with it), then the conversation is more difficult. At the other end of the spectrum, do not nod at everything as others will soon understand that this means nothing and is just a reflex.

- *Attempt to resolve conflict:* Use communication skills to negotiate conflicts instead of leaving things bubbling under the surface until an explosion occurs. In some cases, this might be only about taking discussions offline and trying to settle an argument in private instead of having a huge conflict in front of everybody in a meeting. Acknowledge the good points that the other person has made that you have not considered. Do not create conflicts just for the sake of resolving them.

As an architect, there may be quite a few conflicts because others may believe that the architect lacks the technical knowledge to create an adequate solution. The only way to negotiate this is to explain why the solution is architected the way it is and answer all questions, no matter how technical they might be. This might mean some research is needed before the disagreement can be solved.

- *Empathize:* Try not to be judgmental or biased by preconceived ideas or beliefs and instead view situations and responses from the other person's perspective. If appropriate, offer a personal viewpoint clearly and honestly to avoid confusion. Bear in mind that some subjects might be taboo or too emotionally stressful for others to discuss. This is a great way to avoid or resolve conflicts as most of these occur because of our inability to cross the bridge from the other's end or see the problem from the other's point of view. This will be covered in more detail in Section 5.12, as it has implications outside of communication.

- *Encourage and treat all persons equally:* Try to praise and encourage others, especially when they might be reticent. The goal is to get the most out of communication, so try to make other people feel welcome, wanted, valued, and appreciated in communications. If others know that they are valued, they are much more likely to value the conversation, be more cooperative, and give their best. Encourage open and honest feedback from the receiver to ensure your message is understood to avoid the receiver feeding back what he or she think that the architect wants to hear. In all conversations, treat people equally making sure they get to share their ideas.

   Architects need to eliminate as much as possible the perception that they only listen to the project sponsor and that the developers or infrastructure specialists are less important. If this were the case, it might seriously damage the architect's reputation and others may then withhold (either consciously or subconsciously) information that might be vital.

- *Maintain a positive attitude:* Be friendly, upbeat, and positive with other people. Maintain a positive, cheerful attitude about life: when things do not go according to plan, stay optimistic and learn from mistakes made, including your own. If you smile often and stay cheerful, people are more likely to respond positively to you. Architects with less knowledge who approach an assignment with a smile on their faces are often more successful than others. This is the best way to eliminate tension when communicating with others especially when new to a team (or organization) because a smile will subconsciously make people relax. Remain calm, smile, and try to find out more information, even if this might mean significant changes to the architecture. It is always better to do this early instead of when the document is ready for approval.

- *Do not hold on to bad news:* Communicating early can help avoid misunderstandings and potential conflicts with others. Communicate early and find a way to state the news objectively. It is always a good idea to try to leave emotions out (although it is equally important to pay close attention to others' emotions). Architects should communicate the risks and downside of solutions and always be prepared to list options and recommendations that include next steps and action items. Although bad things happen, communicating on time and being proactive to find a solution will minimize the impact of new problems as they are discovered.

Fortunately, there are quite a few ways to improve your communication skills and to be able to share those wonderful ideas and technical solutions that you are so good at creating. In Section 5.3, I mentioned the AMA, which helps professionals to improve their leadership skills and to cultivate some related performance competencies such as communication. AMA also offers different communication courses designed specifically to meet various communication needs such as:

- Advanced Leadership Communication Strategies: This course is ideal for architects requiring leadership skills by adapting according to changing circumstances.
- The Voice of Leadership: How Leaders Inspire, Influence and Achieve Results: This course helping architects to communicate their solutions and inspire the others to implement them.

Another great way to improve communication and leadership skills is to not only get trained but also to be able to get coached and coach others by becoming a member of a Toastmasters International group. The mission of this organization is to provide a supportive learning environment in which members are encouraged and empowered to develop communication and leadership skills. Many people are not born leaders, are not able to talk in front of a large group of people, or are not able to give and accept feedback. Toastmasters is a great environment to improve soft skills and to be able to practice them and exchange feedback. Their central idea is learning by doing and from obtaining constant feedback. The biggest advantage is that Toastmasters does it in a friendly environment and not at work where the boss can quickly penalize every mistake and, even worse, may not give necessary feedback and coaching. There is a communication and a leadership track and for each of these you can start with a manual and then quickly practice during the club meetings. There are currently more than 15,000 clubs in 135 countries. Be careful in choosing a club that meets your expectations as well as your schedule [9].

As with any other skill, you need to practice, get feedback, adjust, and then continue to improve. The architect needs to make sure that he or she is regularly talking with team members or make sure that the management knows about the issues on the project.

5.5 Presentation Skills

Presentation skills are the skills that you need to deliver effective and engaging presentations to a variety of audiences. Although related to interpersonal communication skills, presentation skills are mostly focused on the method of communicating information by presenting it in front of an audience. For the architect, presentation skills are decisive in convincing the audience about the effectiveness of the architecture solution chosen. This section will discuss the core aspects of a presentation and what techniques an architect should employ to make a presentation successful.

You might ask: What are the main elements re common across all successful presentations? At the core, a powerful presentation has to meet the following criteria: that is, a good presentation must be:

- Clear: Have a logical flow of information and be easy to follow.

- Concise: To the point and specifically targeted for the audience.

- Compelling: Commands the attention and engages the interest of the audience.

Each of these elements has to be reflected in the major stages for any presentation:

- *Groundwork:* Find out as much background information as possible, such as:
  - Who will the audience be?
  - Why is the topic important to the audience?
  - What are the needs of the audience?
  - What is the impact of meeting those needs?
  - Which are the main points for the presentation and do they address the anticipated questions?
  - What is the desired outcome of the presentation?
  - What questions might be asked for which answers should be prepared?

- Although the delivery of the presentation is the most impactful of the three stages, the groundwork can be considered the "worker bee" activity (homework to be done for the presentation) wherein all the relevant information is collected so that it can be used for preparation and delivery.

This information provides important input into your presentation preparation somewhat like the requirements gathering phase in a regular project.

- *Presentation preparation:* This is the most important activity, because this is when you include and tailor the information the audience needs so as to maximize impact. Although no presentation is powerful all by itself (without somebody presenting it), the purpose at this stage is to make sure that all the necessary information is presented and structured in a way to make the most powerful impact on the audience. Each presentation should be organized in three main parts as follows:
  - Opening: This should be captivating and should clearly state the purpose of the presentation, its plan, and a call to action (preparing the audience for what will be needed from them at the end of the presentation).
  - Body: It should have convincing elements (think about how to make the points of the presentation persuasive), benefits statements (what is in it for the audience for each of the key points), and transition (connecting each key point to the next one that follows it). This is when visual aids should be used to increase the amount of information retained by participants. Some examples of such aids include: flipcharts, TV monitors, projectors, and laptop computer presentations. The main purpose of these visual aids is to enhance the verbal content of the presentation by painting a compelling picture (make sure to explain the information properly including symbols and abbreviations). An important part of the body section is the benefit statements, which should be specific for the audience and address the impact of closing the gap between the audience's current and desired situations. All key points should be connected by using appropriate transitions to maintain a logical flow and to keep the audience interested and engaged.
  - Close: This should be the part that summarizes the key points and the action(s) that you want the audience to take as a result of the presentation. You should also transition the focus from content to

questions that the audience might have. It is important to prepare for questions that may be asked prior to the presentation.

- Delivering the presentation: Assuming that all the correct information in the presentation is in the proper format, what is the best way to deliver it? Delivery is what differentiates a good presenter from a bad one. That is because there is a natural tendency to lose the audience's attention, even if the presentation is on an interesting subject.

The purpose of this stage is to get the audience from an uninformed state to a committed to act state. Here are some important tips to make presentations more compelling and successful:

- Body language: This can either enhance or undermine a presentation, no matter how well prepared the speaker might be. If the speaker can inspire trust and credibility, the audience will judge the presentation on more than just the spoken word.

- Eye contact: Just as in one-to-one conversations, eye contact is necessary. Because they are in front of a group, presenters must move around, repeatedly making eye contact with as many members of the audience as possible. Reading attendees' body language is also a great way to observe reactions, gauge understanding and enthusiasm, and look for buy-in.

- Questions at the end of the presentation: Take some time to understand how each question is relevant to the topic, ask the questioner to restate it or rephrase it if needed, and state the response briefly and clearly, making eye contact. Some questions might require an answer after the presentation, but make sure that this is specified clearly and return with that answer later on. If you have to say "I'll get back to you," make sure you do.

In summary, presentation skills are an important tool that an architect should use to foster understanding of his or her solution and to sell that solution to the audience. By the end of the presentation, the audience should buy into it and to be committed to support the architect's solution in front of the other attendees. After all, any project represents a change and the presentation architecture solution is trying to address the awareness (what is the nature of the change, why is the change made, and what is in it for the audience), desire, and knowledge (how to implement the change) stages (refer to the ADKAR model used to manage changes discussed in detail in Section 5.8 [10]), which normally are the most critical stages in a change.

As with any other change, the support of senior management is needed to reinforce the message occasionally, but the first step is to make people understand and want the solution. At times, there will be hostilities because some owners of the system might not agree with the solution, a developer might think it is a total waste of time ("I already have the script done, why do I have to adhere to the enterprise standards?"), or an analyst might be under the impression that the solution does not handle all detailed scenarios. For all of these, the best way to deal with them is to first remain as calm as possible (do not forget to breathe). Then quickly and honestly assess the situation and proceed to:

- Acknowledge the objection; do not overlook it or forget about it. You might forget, but the audience will not. Simply acknowledging a question is a good way to gain trust because the audience sees that you are willing to take it into consideration.

- Decide on the best course of action. Assuming that the research stage was completed thoroughly, go through a quick assessment and decide if this query can be addressed on the spot or whether it should be answered later. When tackling any conflict, the important thing is to be honest and disclose what you do and do not know. This includes the pros and cons for any alternatives that others might suggest. Do not be afraid to change the solution if suggestions from others appear to be valid. Sometimes it is hard, but try to keep your ego out of this and keep calm when weighing the pros and cons. Remember that it is always better to change the solution earlier than later, preferably as soon as you know that change is needed or warranted.

Prepare yourself and rehearse as much as is required to make sure that you eliminate as much risk as you can. Creating a solid presentation but not delivering it effectively and convincingly might jeopardize the entire architectural solution. In the worst case, people would not commit to support the solution once your presentation is delivered.

As with other performance competencies, presentation skills can be improved by choosing the right training and through much practice. There are different courses designed to provide the skills required to create presentations crafted for maximum clarity and impact. There are also workshops with the extra benefit of giving the opportunity to practice by developing presentations, delivering them, and obtaining feedback on ways to improve. A presenter needs to not only be able to speak with conviction and confidence in front of different groups of people but also understand the impact of nonverbal elements and be able to successfully leverage them in each presentation. The previously mentioned AMA also has presentation skills training seminars developed specifically to enhance public speaking style and deliver the message to the architecture stakeholders. Other popular courses include "Strategies for Developing Effective Presentation Skills," giving presentation tips and strategies to present ideas.

Another such course is "Effective Executive Speaking," which makes a great fit for enterprise architects learning to deliver successful presentations to influence executives and shape the direction of the organization.

5.6   Showing Accountability

Because showing accountability is not such a common skill for architects, I would like to start with this definition (Wikipedia):

> the quality or state of being accountable or the willingness, acknowledgment and assumption of responsibility for actions, products, decisions including the administration, governance, and implementation within the scope of the role or employment position and encompassing the obligation to report, explain and be answerable for resulting consequences.

It is quite a lengthy definition. In a nutshell, accountability is about showing ownership for everything one does and manages. This is so important for architects that we should take some time to discuss its main aspects:

- First and foremost, the architect is responsible for his or her own design at whatever level he or she is describing the architecture. What this means is that the architect is responsible for all the architecture decisions that he or she might take during the project as well as for the overall architecture that shows how the solution would satisfy the concerns of the business. The architect does this based on the information available at that time, but to emphasize:
  - The architect should do his or her homework and make sure that he or she collects all the information available on the respective subject including pros and cons for the various options as well as assumptions and constraints that would guide or limit the final solution.
  - All of these items must be properly documented so that anybody else (including the architect) is able to return some time later and understand the reason(s) behind choosing a certain solution. We will discuss more about the various documents produced by the architect in Chapter 7. The important thing to remember about accountability for documents is that the architect should not only accept feedback on such documents but also make sure that such feedback is incorporated into those documents. (Close the loop by making sure that each document covers why the same solution was kept in answer to the person who provided the feedback.) Going back to leadership, this is

an important way to increase the trust and between the architect and the team or the account manager that sponsors the initiative.

- The architect is also responsible for providing various views for his or her solution that would show how this is actually satisfying each stakeholder's concerns.

- The architect is responsible for the architecture governance for the whole life cycle of the initiative. As an example, the software (application) architect should provide the governance during the whole software development life cycle. Likewise, the enterprise architect should provide the governance from the moment of receiving the request for architecture work, going through the definition of the architecture vision, architecture current/target state and roadmap for the architecture domains, and implementation strategy, and finishing with the implementation of all the work packages that needed to be implemented.

Once the enterprise architect defines the architecture vision for the enterprise, he or she owns that vision (normally, together with the chief information officer). As a personal example, in one of the major financial companies in Canada, one that was formed through acquisitions, there were a lot of redundant systems with very little integration. As part of the enterprise architecture team, I not only defined a reference architecture, target state, and roadmap for the integration platform, but I was also accountable for its implementation across the enterprise, guiding the solution architects working on the projects, which made the vision come to life, adjusting where necessary but in the end working with the project teams to get the job done.

Being accountable means something more than just being responsible for everything that the architect *is supposed to do*; it also means helping others become accountable as well. Thus, accountability means demonstrating a proactive view: seeing a problem and handling it, it shows that you know where you stand, that you measure your progress, and that you do not give up. It also means that you help others to understand what accountability means. It means helping them to go from denial or blame ("I didn't do this, but I know who did") to the point where governance becomes more of a boring list of checkpoints because they are on top of things. It is about learning how to help people feel more responsible and not "micro-leading" everything they do to make sure it is OK. It is more than just meeting commitments and solving problems. It is about solving the problem first and then helping others to find the root cause and preventing a recurrence.

The others on the project expect this high level of accountability and leadership from the architect and this is one of the ways to show and increase that trust. Most of the time, the success of the project from the technical point of view will depend on how accountable the architect is to the solution, to the project, and to the team. The team will see the architect as a technical role model. As a result, if the architect only delivers the architectural solution and then does not follow up, does not lead people, and *own* the solution up to its implementation, the team will also just do their jobs without caring too much about the project overall. As with the other performance competencies, there are also training courses that are mostly geared towards creating a culture of accountability. One of the best courses is the Fierce Accountability Training Program (for more information, please visit: http://www.fierceinc.com/programs/accountability), which takes as its objective fostering accountability within teams and throughout organizations.

5.7　Planning

Most of the time, the concepts of being an architect and being a project manager (PM) seem to be at the opposite ends of the IT spectrum. However, an architect has to:

- Have good abilities to plan and organize activities for themselves as well as for the group. The project manager would not be able to identify those technical activities or validate ones coming from the developers, infrastructure specialists, or testers. The architect is the one who should collaborate with the PM to make sure these activities are captured in the project plan. This includes preparation for contingencies, itemizing and prioritizing tasks for oneself and others, managing dependencies and architectural risks to ensure that the deadlines are met, and being able to proactively define alternatives to deal with slippage on the project. Basically, architects interact with project managers to provide them with technical domain knowledge and problem-solving skills and to identify and resolve issues that might occur. Although a project manager usually is the person actually performing the project planning, the architect should also be aware of that planning process including: identifying project stakeholders and their concerns [which is also a step in one of the most used enterprise architecture frameworks, The Open Group Architecture Framework (TOGAF)], identifying project deliverables (usually by tailoring them from a project methodology), creating the list of tasks (the architect should have accountability for all the technical tasks), and adjusting the plan whenever necessary.

- Show ownership of the project plan and make sure the architect fully understands the tasks and how each change in assumptions and constraints would also influence the plan. For the enterprise architect, it also means owning the project that would define the work packages (projects) that would shape the enterprise according to the business strategies. This requires strong planning capabilities as they need to be able to handle multiple iterations or multiple concurrent projects and to be able to coordinate the changes introduced by them.

- Be able to coach and mentor others on how to better plan the tasks and come up with best practices and standards for planning technical tasks. The architect is one of the senior people in the team, so the other team members expect them to be a role model. Other team members also expect architects to be able to help them with management of tasks, including how to prioritize them better as well as how to juggle multiple tasks simultaneously. Showing ownership of the project plan also means helping others find a way to minimize the impact on deadlines because of various changes that might occur.

If the architect is supposed to know so much about project management and planning, why not actually be an architect and a project manager? In most startup and small companies, the architect is also expected to manage the project as well. This is the point where the boundaries between different IT jobs get blurred as the person hired as architect is expected to also do a little bit of project manager and be *a little bit* of a developer as well. To be absolutely clear, although planning is an important skill for architects, it does not make them project managers as well.

There is a reason for the separation of the jobs in IT, which is the same as why we had job specialization in the ancient world: making sure society improves by having certain individuals doing specialized tasks. The project manager and the architect should understand that they should not compete but they should complement one another. The architect should understand that the project manager's task of managing the project will actually relieve him or her of the tedious job of putting together and adjusting the project budget, creating project baselines and charters, and sitting in the meetings with the project sponsors. The project manager should understand that having a good architect on his or her project is a great relief as the architect will own the technical solution, be in constant communication with the project team members, and provide valuable technical advice to the project manager. As with architecture certifications, there is a Project Management Institute (PMI) that validates and certifies your knowledge based on rigorous standards designed to meet the needs of current organizations. This institute provides a number of certifications including the PMP (the Project Management Professional; the gold standard of the project management certifications, validating the ability to lead and direct project and teams). Some architects who perform the dual role of project manager and architect decide to take this training and obtain this certification, which is sometimes paid for by the organization if the architect will play the role of the project manager [11].

5.8 Change Management

Change management is defined as an approach to shifting or transitioning individuals, teams, and organizations from a current state to a desired future state. This section will describe change management skills used by the architect to get the project team through an architectural change. Together with other things in IT, architecture is by definition a change. Each architectural initiative or any kind of architectural changes are going to affect the shape of the whole enterprise, of a collection of systems, or the interface between applications or devices.

Much as with planning, an architect is supposed to know how to manage change, although most of the time, he or she will not act as a change manager. It is important for an architect to identify changes and to be able to apply a certain framework to better understand them so they can help change managers to drive the project and recognize any roadblocks that might affect the implementation of the architecture solution. The architect is the one who can take ownership for the overall change design, development, and successful implementation. This change might be at multiple levels including people (how will the architecture change influence the organization of the company), business processes and functions (business architecture), data or application changes (information architecture), or changes in the infrastructure (infrastructure architecture). In line with the change management definition, the architecture (at the various levels starting with enterprise architecture and going down to application/infrastructure architecture) is about defining the current state, the required target state, the gap between the two, and a roadmap of how this change will be implemented. Change is always about the current, future, and transition states, in the same way that architecture is about defining current and target (future) state and the (eventual) transition states that it has to go through. From that perspective, an architect must be the champion of the (architectural) change and be able to explain the benefits of the target state to alleviate any worries. Resistance and being worried are natural human reactions to change, so an architect should expect resistance to any architectural changes (and be worried if none comes). This also comes back to the leadership style, but threatening people with the change or trying to enforce it by getting help from the sponsor is never a good idea as the team will lose trust.

As with any problem, there are various tools in the form of change management frameworks that provide structure required in managing change. One of the most popular change management frameworks is the ADKAR model (which stands for Awareness, Desire, Knowledge, Ability, Reinforcement); it is a goal-oriented change management model that allows change management teams to focus their activities on specific results. The ADKAR model is able to identify why changes are not working and help you to take the necessary steps to make any change successful. The ADKAR model (please refer to Figure 5.1) also allows breaking down a change to assess it for each of these five building blocks:

- *Awareness:* Understanding what is the nature of the change and the need for change.

- *Desire:* Understand what is in it for me (each of the participants in the change) and decide whether to participate and support to change.

- *Knowledge:* Understand how to change.

- *Ability:* Have the capability to implement the change.

- *Reinforcement:* Actions that increase the likelihood that the change will be continued.

The architect's role is to lead the other members of the team through the architectural change by doing the following:

- *Getting support* from the project sponsor because he or she is the one who can do a lot to push the change through. One of the main causes for most failed projects or change initiatives is a lack of support from the sponsor. Once the architecture solution is defined and there is a clear idea of what changes are necessary, build a coalition with the sponsor and determine the main messages for the team members.

- *Communicate* the architectural change properly to everybody in the team including what are the benefits of the change as well as the risks and why we need this change. For example, when introducing a new application, system, or interface, it is imperative to explain why it will be introduced because most people will be happy with the current state and will see no reason to move to a new way of doing things. If members of the team lack awareness, they often start to resist change and this is definitely something to avoid.

- *Advocate* the architectural change because otherwise team members will not view it as important and they might start to push for the old way of doing things. The architect should not only know the architecture inside out but should also be the champion of the architecture, making sure to

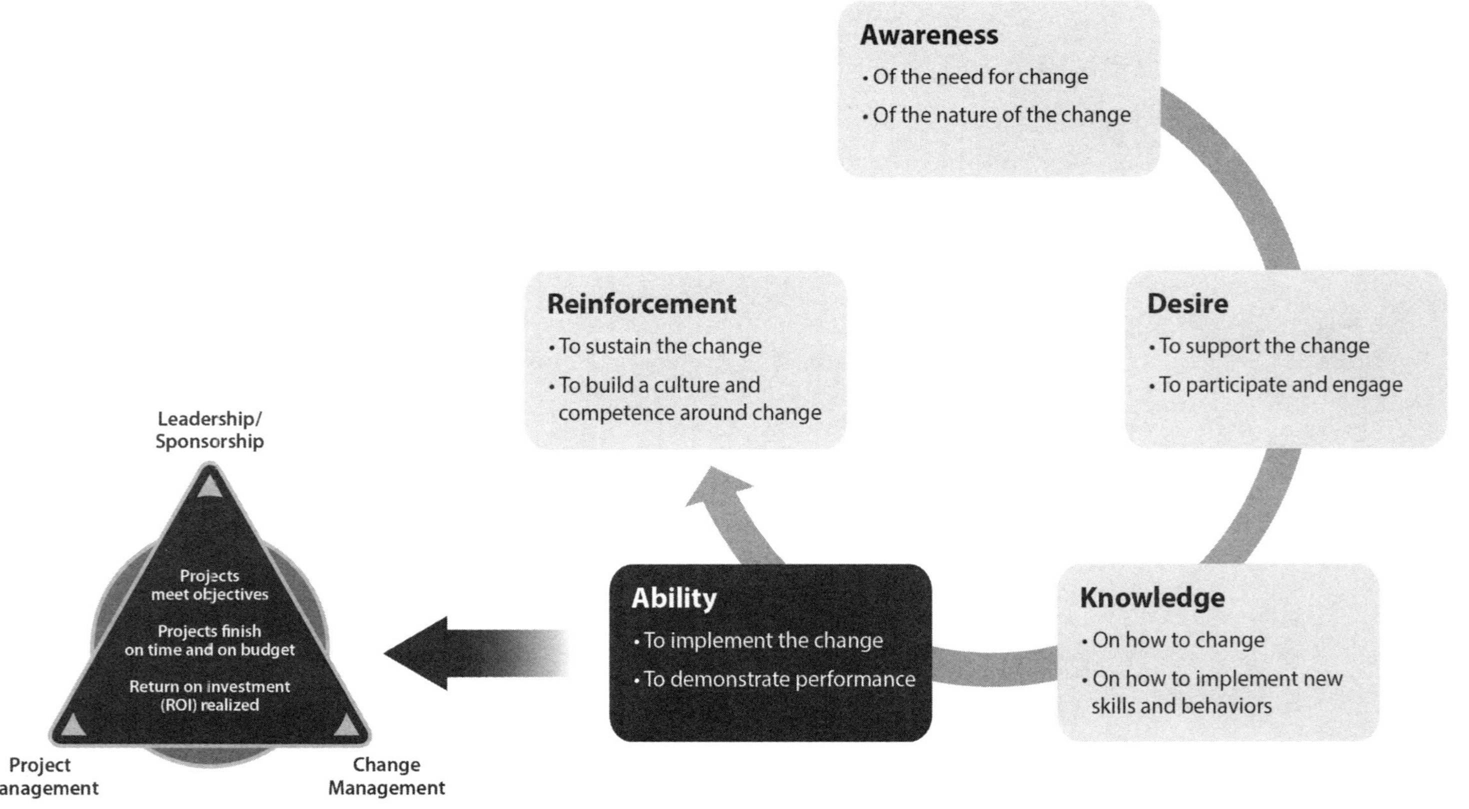

Figure 5.1   ADKAR model change management framework. (Source: ADKAR model. © Prosci, Inc. All rights reserved.)

build awareness of a need for change and making sure that team members have the knowledge and desire to work through this architectural change. Being an architecture champion means constantly reinforcing the message that change is coming, and change is for the better.

- *Coach* them through the change. Do not forget that most of the time the architect is one of the senior people on the team. Thus, it is part of the architect's job to coach the others through the change because they know the most about changes and also they know the most (at a high level) about this particular architectural change.

- *Be the connection* between the project team and the sponsor and management. The architect needs to start building an alliance with the sponsor and making sure to bring an accurate view of how the architecture change is perceived by project team members. If there is any kind of resistance, it needs to be discussed so that, together with the sponsor, the root cause can be identified and an effective solution found.

In summary, architecture itself is a change for the organization. Whether it is a big change represented by enterprise architecture (revamping the enterprise business and systems based on the business strategy) or some small change in the application architecture (like application interface changes because one of the application has been redesigned), the architect has to know how to effectively manage the change or work with others (change managers, account managers, project managers, and so forth) to manage the change.

5.9 Stakeholder Management

Stakeholder management is an all-important skill that an architect can successfully use on the project to ensure its success by getting support from the stakeholders. Having a well-designed architecture is only the half the battle toward a successful project implementation. The other half has to take into account dealing with the human resources on the project. For this, the most important stakeholders must be identified early in the project and their input must be used to shape the architecture to ensure their later support and the validity of the architecture model. Frequent communication with the stakeholders is important to make them understand the architecture process and the benefits of the architecture as well as to anticipate and address any likely reactions to the models. The ability to build into the plan the actions that will be needed to capitalize on a positive reaction while avoiding or addressing any negative reactions is necessary.

One of the first steps towards stakeholder management is understanding the culture of the organization. As opposed to the organization's values and strategies, organizational culture is not posted on the internal Web site or published in a booklet. The culture is the set of shared beliefs, values, and assumptions held by the people in the organization, embedded in the organizational practices and vetted (amplified) by the behaviors of their leaders. Understanding the culture is not an easy task, but an architect needs to put a lot of effort into this as otherwise people will resist the architectural solution (or vision for the enterprise architecture) being put into place and the whole initiative might wind up on the brink of failure.

In many organizations, the culture is not obvious and the beliefs and values held by the people working there are quite different from the values publicly promoted by the company. Sometimes architectural changes need to take into consideration the internal culture (especially at the enterprise or business architecture level). The architects need to be aware of *how things get done* every day in the company and proactively think about the effect of this to the architecture solution. Architects are leaders in the company (and can amplify some of the cultural behaviors), so they might also be involved in efforts to change the organizational culture.

An enterprise architect could recommend changes that might affect the organizational culture. First, the culture must be understood to recommend changes to it or participate in a cultural change. Jack Welch's leadership secrets book is full of examples of things that did not work that he had to change to help General Electric (GE) survive [12]. Some cultures are totally resistant to architecture in general, and executives trying to change things may find themselves embroiled in a constant battle to push for architectural changes that might be a good benefit for the company in the long run. This is why it is important to understand the company culture. Sometimes it might be about having one-on-one meetings with the important stakeholders and walking them through the solution trying to proactively show how the solution addresses their concerns or sometimes it might be about the feedback received from them. Whenever you get any kind of feedback on the architectural solution or the decision being proposed, analyze it and either incorporate it (even partially) in the solution, or (very important) if it is not going to be incorporated, circle back to them and let them know why those changes to the solution based on their feedback will not be implemented.

There are many ways to manage stakeholders. Some of the architecture frameworks devote entire chapters to talking about stakeholder management. This is beyond the scope of this book, but as an example, the open group architecture frame work (TOGAF) is a framework for enterprise architecture defines a few simple steps (please refer to Figure 5.2):

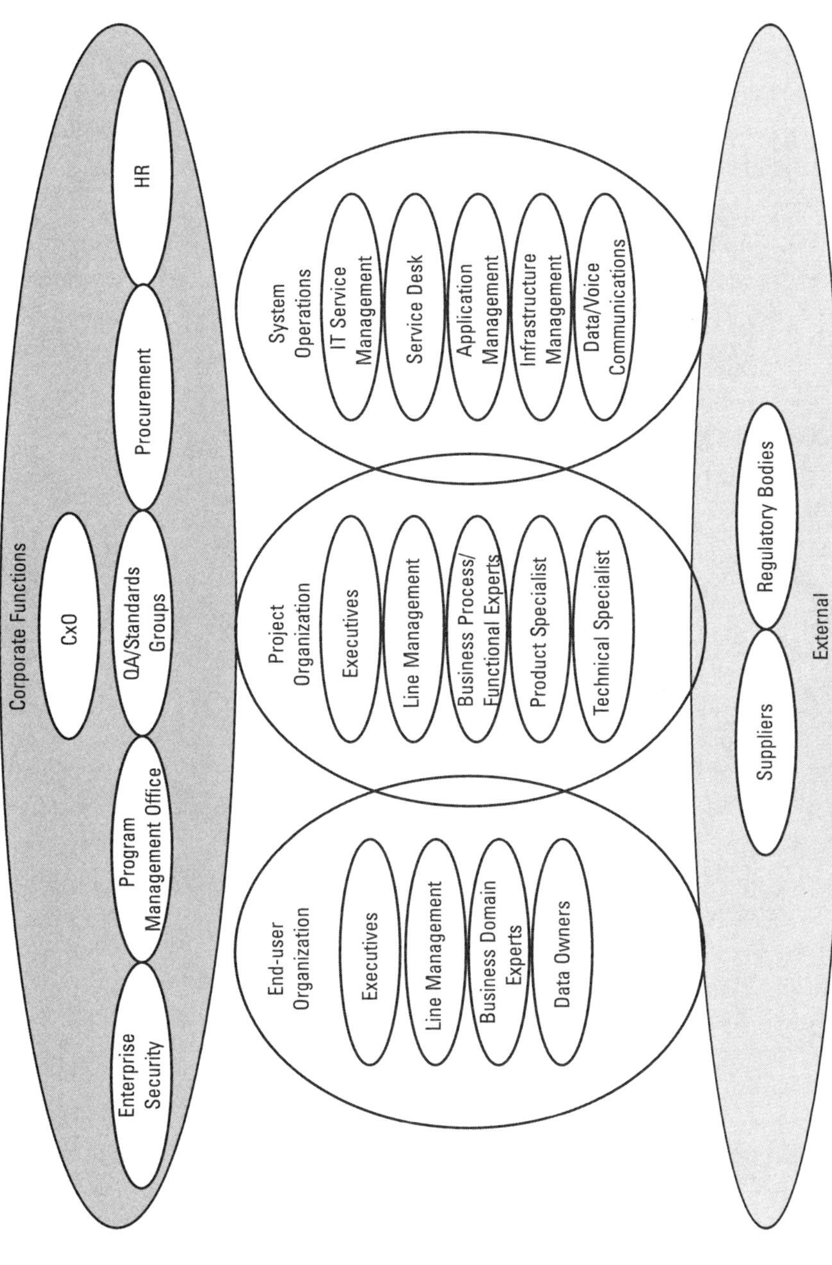

Figure 5.2 Stakeholder categories. (Source: TOGAF® 9.1.)

- Identify the stakeholders: Think of all the people who are affected by it, who have influence or power over it, or who have an interest in its successful or unsuccessful conclusion. This might include senior executives, project organization roles, client organization roles, system developers, alliance partners, suppliers, IT operations, and customers.

- Classify stakeholder positions: Get an understanding of the most important stakeholders and assess the readiness of each stakeholder to behave in a supportive manner.

- Determine stakeholder management approach: Understand which stakeholders are expected to be blockers or critics and which stakeholders are likely to be advocates and supporters of the initiative.

One of the important points in getting the support from stakeholders is coming up with the right architectural views that show how their concerns are addressed in the architectural solution and what its benefits are from their point of view (what is in it for them). For an enterprise architect, this might be about selecting the right views to make sure the chief information officer and the senior IT executives buy into the proposed enterprise architecture solution, while for the business architect it might be about selecting the views to show how the proposed business architecture would satisfy the business strategy.

In summary, understanding the organizational culture and the stakeholder management in conjunction with strong sponsorship are important steps towards the successful implementation of the architectural solution.

5.10 Consulting Skills

Consulting skills can be defined as [13]: "understanding the results that stakeholders desire from a process and providing insight into how efficiently and effectively those results can be achieved." The architects must demonstrate strong consulting skills to successfully implement their initiatives. Consulting represents a set of skills that include advisory skills, consensus building, effective client relationship, effective communication, and, in general, provision of excellent client service.

There is a general preconception that only the external consultants or the employees of the consulting companies should have consulting skills; this is not true. Every IT professional, whether an internal or external consultant to an organization, needs to develop consulting skills; this is especially true for architects. Consulting skills will help them partner with their clients so that they can better empathize with the client. In this way, they help to build a relationship with their clients and take ownership in solutions to their problems. The other common preconception is that consulting skills (for the internal consultant) are only needed when talking with the business. The reality is there is always at least a subset of the consulting skills that will be used in most of your interactions with the stakeholders or the members of the project team.

Many books discuss in detail how to successfully provide consulting services, and most of the information applies to the architect role. The important thing to understand is to adopt a consulting mindset by:

- *Dressing for success:* It has always been said that the first impression matters, although nowadays this is not as strict as it used to be. We should always assess the situation and dress accordingly, as sometimes wearing a suit for a normal team meeting might not be the right call but also wearing jeans for a presentation with a C-level audience might not help you to deliver your architectural message. This is actually about more than just the clothes that you wear because it includes everything that might affect your image, such as acting in a professional manner. Wearing a suit every day does not make up for acting unprofessionally.

- *Motivations and feelings:* Try to understand what the client's motivations are, what is driving them, what the client thinks the problem is, and what he or she expects as the outcome. Keep in mind that the client might be internal (the business, account managers, and so forth). Be aware of the client's feelings and listen to the client first to understand the problem (which might have a human component). Emotional intelligence will be discussed later in this chapter, but one component of emotional intelligence that is essential for the consulting mindset is to always try to bridge the gap (between you and your client) from their side.

- *Perceptions and politics:* As mentioned in previous chapters, there might be some preconceptions that the architect knows everything. There are subject matter experts in the company who know more and the sponsor also knows much more about his or her business than the architect. Being humble and showing respect for the knowledge that they have is always a good way to start the relationship. Sometimes the architect's role is about understanding what they know and what they would like to achieve. Then the architect must translate that knowledge and aspiration into an architectural solution that should obtain support from major stakeholders and will add credibility to what the client was saying all along (a great way to increase trust). Understanding the politics (including who does not like who or what) is also important for consultants because they need to understand relationships at work before they speak and try to stay neutral on any conflicts between teams or individuals.

This relates to understanding the organizational culture as mentioned previously.

Many consulting models have been proposed over the years to address such matters. One of the most useful such models was proposed by Peter Block in his book *Flawless Consulting* [14]. It includes five phases of consulting that go from the initial conversation with the client discussing the problem and the expectations, through data collection and the diagnosis phase, and ending with the client's decision to go ahead with the implementation of the proposed solution architecture or to scrap the project (please refer to Figure 5.3). One of the important elements in the first (entry and contract) phase is the statement of work, which clearly states the work in terms of scope (what will be done) as well as which key performance indicators will be used to measure success. This consulting model applies very well to the work done by enterprise architects (as described in architecture frameworks and methodologies like TOGAF) who are usually brought in to provide recommendations for a specific problem. After the first phase (when they clearly document the expectations), they would iterate through the data collection, diagnosis, and report phases. They will put together the high-level architecture vision and get the go-ahead from the sponsor (first iteration) and then they would delve into the business, data, application, and technology architectures, define the gaps and roadmaps, and develop a list of work packages to be approved by the client (sponsor). Once they get a second go-ahead, they would start with implementation and governance. For other types of architecture, the consulting model is still valuable but may require some tailoring or customization.

An application architect, for example, could still use his consulting skills but the engagement model might be different because most of the time they

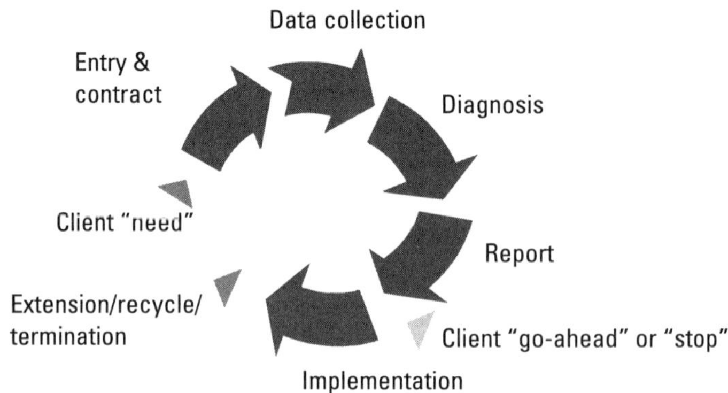

Figure 5.3 Flawless consulting model. (Source: *Flawless Consulting: A Guide to Getting Your Expertise Used.*)

have the go-ahead from the sponsor (in the form of the project to which they have been assigned). Instead, they work on stating expectations, collecting data, solving the problem (diagnosis), and putting together an architecture solution. In this case, the client go-ahead often represents approval of such a solution by the sponsor. For the business architect, the tailored consulting model might be quite similar to the one for the enterprise architect but with the scope limited to the business domain architecture.

Once the most appropriate consulting model is established for the specific architectural role and initiative, some of skills that can help with a successful consulting engagement include:

- *Communication:* This is definitely an important element, but communication was covered in detail earlier in this chapter, so no further details will be given here. Honestly state what can and what cannot be done and seek information about real, underlying needs of the client, beyond those expressed initially.

- *Negotiation:* The ability to influence another party including the client and any member of the project team. Negotiation is all about getting to a win-win situation, but the hard part is how to get there. The architects need to focus on the result and sometimes get the best deal knowing that they might have to change their solution to reflect another party's opinion. Obviously, trust is an important element in this process: the better the trust, the faster the deal will get done. Keep in mind what architectural solution would still be acceptable that would not jeopardize the target state of the system or would achieve the business strategy for the enterprise. Be aware of any internal conflicts and animosities that might affect the final solution. These might turn out to be a huge roadblock (if unknown before the meeting), but they can also be an excellent way to improve trust (if resolved properly). At the end, commitment to the final solution (which might be different from the one initially proposed) is needed from all parties involved.

- *Visualization/presentation:* An architect must be able to draw diagrams and this becomes even more important as a consultant. The early discussions with the clients are best summarized in a diagram. It is recommended to paraphrase when listening so the other party understands that they have been heard but being able to not only repeat the words but also visualize it in a diagram is a great way to start a relationship. An architect must understand the modeling language and create visual representations for the different abstractions and viewpoints. In discussions with technical staff, go into detail about the application or infrastructure components but in a meeting with business people use a viewpoint that shows the business processes, functions, and roles and what application services would automate them. Such a visualization can also be combined with the presentation skills discussed earlier in this chapter to increase success. Combine the presentation skills tips with technical knowledge to create the visual representations on the spot and show the client an understanding of his or her problem. Provide realistic options and solutions, taking into consideration all the known constraints and assumptions. Figure 5.4 depicts an example of a business viewpoint that describes business capabilities and functions, as well as the application services that realize them.

- *Planning:* Although the architect is not responsible for producing and maintaining the overall project plan, he or she should assist the project manager in planning large projects starting with the scope definition and continuing with various technical tasks. The architect must understand the basic concepts of managing a project so that he or she can come up with a general description of how the project will be executed. Because each project is unique, it is important to understand the roles and responsibilities for each of the team members as well as the stakeholders or the steering committee that will guide the project and provide approvals for the various deliverables. The architect is usually engaged either before the project starts or during the initial stages of the project so that he or she invariably knows the business or technical problem that the project is trying to solve as well as the high-level design options that are available. It is helpful to document those project constraints and related assumptions because that will help with the scope definition and the initial estimation of high-level project tasks. The architect will also validate assumptions during the project so that at the end of the architecture phase, all (or most of) the assumptions will be checked.

An architect has to assist the project manager in identifying the skills required for the other members of the team, for example, identifying that a certain project requires a mainframe developer or a network specialist. The project manager will follow up and request this kind of skill set from the internal resource pool. He or she will have to determine the initial allocation (based on the estimation) and then fight for resources as sometimes the availability might be a problem.

An enterprise architect guiding the implementation of the work packages previously identified will also have to manage the relative priority of these projects and their dependencies. Any other type of architect will work at one point in time on multiple projects, so time management becomes very important. As

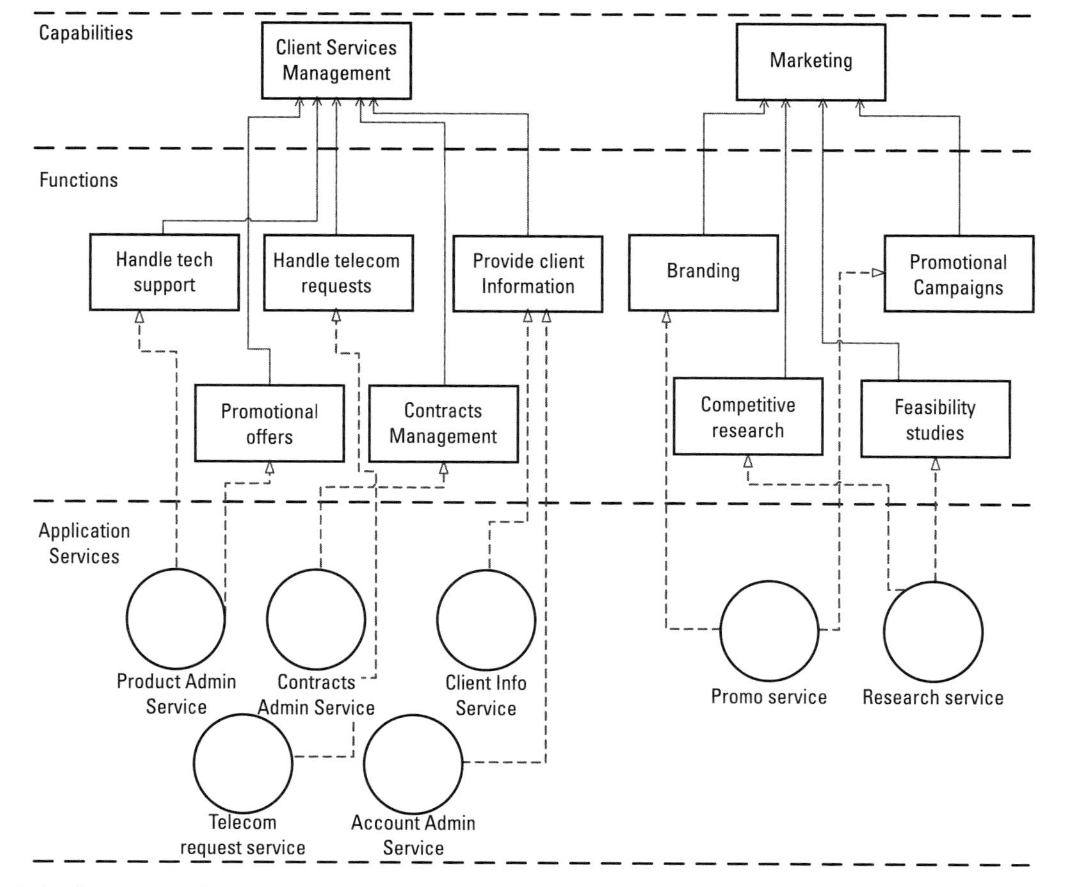

Figure 5.4 Example of visual representation of the business capabilities, functions, and application services from a financial enterprise.

mentioned previously, an architect might have to play the project manager role in some small projects, so it helps to be familiar with some project management techniques. There are quite a few good books on project management as a discipline, which is beyond the scope of this title.

Earlier, we covered consulting skills that an architect can use daily. There is a distinction between internal and external projects and the type of consulting skills used. Although consulting skills are helpful in both cases, contractors need to have even better skills than permanent staff as they are mostly brought in just for a specific assignment. They need to quickly understand the culture and expectations and be able to produce results in relatively short time. As with any other skills, there are different levels of expertise and contractors need to be experts in communication and negotiation.

As mentioned for the other performance competencies, there many training courses available to help an architect develop strong consulting skills. The Association for Talent Development (https://www.td.org/Education/Programs/Consulting-Skills-Certificate) has a Consulting Skills training course that defines a model, describes techniques and tools for consulting, and teaches IT professionals to provide feedback in the most appropriate ways to management. The AMA has an Internal Consulting Skills Workshop that prepares attendees to diagnose a situation within an organizational context, develop a problem solving approach to addressing various consulting challenges, and learn critical skills that a consultant should have. The workshop recognizes that almost everyone needs some degree of consulting skills and even if they are not explicitly specified in job descriptions, good consulting skills can help in almost any role.

5.11 Selling Skills

The architect as a salesman? The salesperson skill might seem a little odd for an architect, but it is not. In some books, selling skills might be included in the consulting skill section, but this book discusses them separately. That is because the better the salesperson, the better the chances for a smooth implementation of an architecture solution.

This raises interesting question: Why does the solution need to be sold to the client? As covered previously, the role of the architect may not seem necessary to everybody in the company. Even after years with the same organization, people still sometimes question why an architect is needed on a project. They wonder what value the architect has. A developer's role is clear: he or she produces code. A tester tests the code the developer produces. What does the architect have to bring to the table?

In many ways, in any kind of job, some kind of selling is required. Sometimes this might be a concrete sale such as selling a car, but at other times, as in the case of an architect, it is about selling a service. Nobody knows you (and your job) better than you, so you should start learning the best way to sell your services. This is valid for all jobs in the IT industry, but for an architect, selling is even more important because others do not always have a clear idea of what value he or she adds to projects. What could be a better way to sell than involving stakeholders in coming up with a solution? As a general rule, people will not complain about a certain decision or way of implementation *when they were involved in making that decision.* Subconsciously, they understand that being part of the solution or decision means that the responsibility for a successful implementation lies on their shoulders as well.

For example, selling skills are very important for an enterprise architect. He or she is the person who has to connect two internal worlds that exist in each company: business and IT. He or she should take decisions and strategies formulated across the entire business, create a business capability map, and see how IT can act as an enabler for these strategies and automate some of those capabilities.

This is just the first part of the job; the second part of the job is about selling:

- How will the set of work packages proposed make life easier?

- How can these capability maps be used to drive strategic discussions on what capabilities are most important to the business, where to expect the most change, and what the most important outcomes of their investments (both business and IT) should be?

- How can capability maps be used to coordinate planning around business and IT change?

- How can the solution create a better coordinated and more consistent execution of the chief executive officer's direction? How can it provide a way to measure that IT projects actually deliver business value?

This applies to all other kinds of architects, although, in general, for lower-level architectures, the sales effort might be somewhat less. An application architect needs to sell his or her architecture as well, but this means explaining why he or she picked certain frameworks or patterns and what benefits a certain option offers over the others. A systems architect would need to sell why his or her solution is better for the system in the long run and how the application and infrastructure architecture offer the best mix considering target and transition states.

Depending on the audience, the message might have to be adjusted and translated from technical terms into easily understandable language. When enterprise architecture is sold to the chief information officer or major stakeholders, those presentations might be framed around cost reduction, entering new markets or increasing market share. Those people do not need to know about what kind of architecture framework will be used (and they do not care), but they would like to see how this new enterprise architecture will grant them more control to steer the company and to make more money for shareholders. For a technical audience, the high-level architecture might be translated into how new architecture changes will affect their system or application and what benefits that confers in the long run.

For all the kinds of architecture, the best way to sell that architecture is to show or demonstrate its value. The project sponsors would love to see some savings, so show them the money. The only problem is that most of the time savings are not immediate. Unlike the software development life cycle (SDLC), which spans one project, the architecture life cycle spans multiple projects. Changing the architecture from a current to a target state usually requires more than just one project. The benefits of the target state (including financial savings) will not materialize until a series of projects have been completed, each adding incrementally to the current state architecture. The difficulty in this case is selling the value of the changes in the first project when benefits will not actually materialize until the end of roadmap implementation (when all projects are built). It is a sometimes tricky task, but the solution is always to show the business value that is associated with each project and then to put each decision in the larger context of a roadmap for implementation.

As the architect, you should always keep in mind a picture of how to get from the current to the target state and what projects (or work packages) and transition architectures (transition states) are required to get there (see Figure 5.5). This is the elevator pitch that the architect should be ready to give at any time, to any audience.

Some might say that these selling skills are actually a combination of other skills already mentioned such as planning and good communication. Selling skills are covered separately here because the presence or absence of these skills makes a big difference for any architect. Architects must be able to successfully sell their architectures: the best way they can do this is by listening actively to stakeholders so that they are sure that their solutions meet expectations. Architects must be able to build relationships with stakeholders and establish rapport. Then they can completely eliminate the initial negative reaction that a manager might have if the architect is just a new person to him or her. For some people, selling comes naturally, but some IT professionals will need training to improve their ability to sell their solutions.

Many sales training courses are geared towards salespeople working in sales departments. These are not useful for architects. Although some elements in these courses are good, taking one would be like taking a systems administration course just to learn how to operate a computer. The AMA offers a course

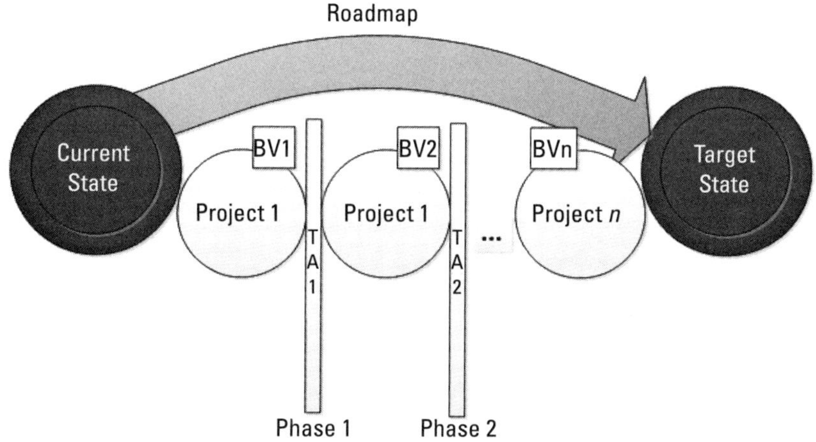

Figure 5.5 Current versus target architecture states.

called "Selling Ideas" that is a much better fit. This course is about getting buy-in and arguing a case effectively, defending a viewpoint, conveying enthusiasm, and developing that personal charisma or influence that helps to push things forward and convinces others that the presented solution is the right one. The course also helps attendees to understand how to conduct effective face-to-face meetings, anticipating conditions that might persuade stakeholders, team members, and decision-makers of all kinds.

5.12 Emotional Intelligence Skills

Simply put, emotional intelligence is defined as the ability to perceive, control, and evaluate emotions. A more complex definition comes from Peter Salovey and John D. Mayer as [15]:

> the subset of social intelligence that involves the ability to monitor one's own and others' feelings and emotions, to discriminate among them and to use this information to guide one's thinking and actions.

What does emotional intelligence have to do with the architect's job? We think of ourselves as cool technical people instead of as people who specialize in psychotherapy. Although emotional intelligence is not on the frontline of required skills, an architect needs to understand emotional intelligence and be able to apply it in day-to-day life. An architect's work depends on the information that they get from project team members and the way that they provide information to stakeholders. Taking into account people's emotions and being aware of the impact that certain decisions or conversations cause are necessary skills for any good all-around architect. Architects must not only be able to create technical solutions but also be able to communicate them in the right way.

As with other skills and competencies, this section does not go into detail (there are plenty of books that cover emotional intelligence). Instead, I will summarize some basics:

- *Self-awareness* is the base for emotional intelligence, a correction mechanism that helps us to be conscious of how we look to others. To understand those others, we must first understand ourselves and be aware of our emotions and how they impact our own behavior. One of the important behaviors for an architect is a strong belief in one's ability and a willingness to stand up for one's ideas (translation: architecture solutions). Combined with a belief of being able to influence an outcome, with control over one's impulses (staying calm under pressure and remaining able to overcome difficult emotions), and with comfort with uncertainty, these characteristics make a good start when creating successful relationships. Another important thing to remember is that people can feel the attitudes directed towards them. Being authentic and disclosing as much information as possible but also admitting when you are wrong are a great way to boost the trust that others are willing to grant you.

- *Effective emotional management is the next step in developing emotional intelligence.* One might think that these are all good concepts, but how are they to be applied? Let us say that you become aware of the impact of your behavior and you want to change it so that the next time you deal with someone who does not care about the project or constantly irritates you, that person will see a new you. The best way to deal with these situations is to first stop (disengage the trigger) and then take a deep breath so that you give your cognitive system a chance to kick in and outrun the amygdala, the part of the brain that is usually faster and responsible for getting us into fights. Once the cognitive system takes over, you are in good hands. You can start to ask questions like: Is this a real or perceived threat and how big of a threat is this after all? Eventually, you will develop a deeper understanding of the problem and how to react to it in the best possible way. That is also using emotional intelligence to your advantage!

- *Be aware of others' feelings.* Without knowing or accepting it, we live in a constant emotional state. Having emotional intelligence skills is not only about being aware of your emotions and finding a way to better react the various situations. It is also about being conscious of others' emotions. Basically, this is about getting to a common understanding of the problem. Because we usually only see our side of the problem, it is not easy to actually build the bridge (to create the communication channel that would allow the common understanding of the problem) and get to understand their side of that problem. First, you have to mind the gap and acknowledge it. Then you should start listening to their concerns, intentions, and driving needs. Once you can identify and label these things, you can help your interlocutor recognize what impact they have on you and then look for solutions together. Something to keep in mind is that a question engages the cognitive system, but a statement triggers the amygdala and destroys the bridge that you are trying to build (see Figure 5.6).

An architect needs to be aware of what people mean and not what they say and also to acknowledge their issues. As an architect, the best rule to apply is to first understand others and then help them to feel understood. When an architectural solution is put together, address the team's need for understanding and make sure to always give them an answer to the (sometimes unasked) question: Why did you do it this way? Never respond to what has been said, but try to understand what they actually mean and acknowledge that by paraphrasing. Especially for topics that might be controversial, smooth things over by trying to find out what their fears are or what perceived threats there are that would keep participants from opening up. Likewise, it is also important to understand

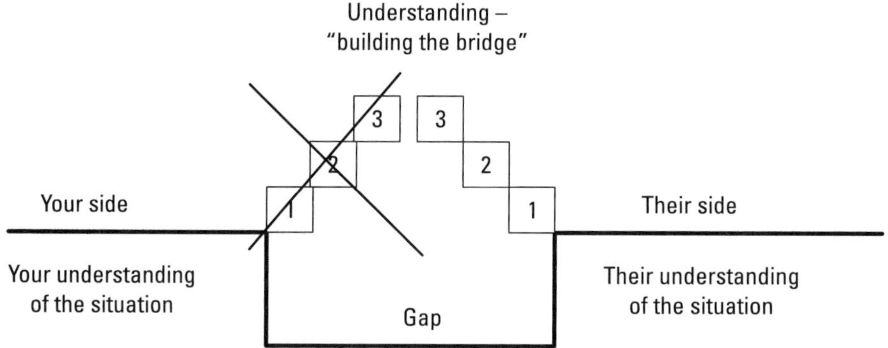

Figure 5.6 Get to a common understanding (building the bridge from their side).

what actual emotional needs they have (the need to be respected, valued, or accepted in the team or in the organization, the need to have a successful job and pay and have power and control, and so forth). Once their emotional needs are understood, try to address them before the meeting (in a face-to-face conversation). This clears the way for the architecture documents that need to be accepted not only by the major stakeholders but by the people who will implement the solution (remember to sell the architecture).

In summary, increasing emotional intelligence abilities will most definitely help in becoming a better architect. Ultimately, these skills will also enhance one's personal leadership abilities. Many books cover this subject in detail; some of the best are listed in the references. Likewise, there are also useful courses on emotional intelligence. Not surprisingly, many studies link having a good emotional intelligence with being an excellent leader, increasing sales revenues, and retaining clients.

Of course, a strong body of technical skills and knowledge lays the foundation for a good IT architecture career (please refer to Figure 5.7). Combining higher emotional IQ, solid technical skills, and emotional competencies makes for the best architects.

5.13   Facilitation

As an architect, working in a team or a group of people is part of the day-to-day job. Members of the team come with different expectations and experiences, so sometimes a discussion must be facilitated to make the most of it. Such facilitation requires certain skills and personal attributes that are not necessarily inborn, but that can be learned. Although the architect's job is not only about facilitation, there are some situations when a discussion needs to be facilitated

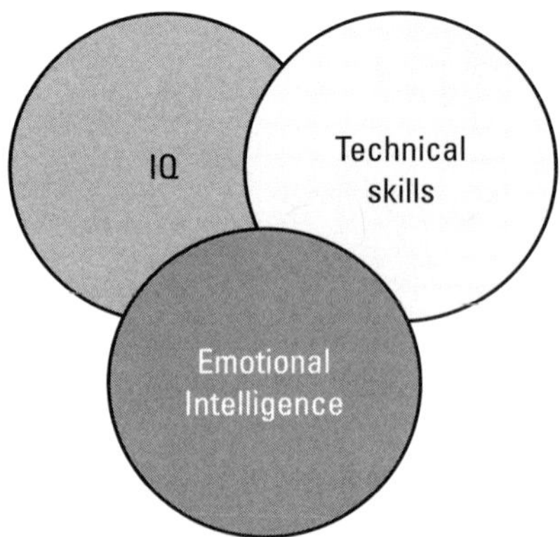

Figure 5.7   Combination of technical skills, IQ, and emotional intelligence.

that include conducting a brainstorming session to find options for a certain architectural problem; to get members of the team more involved and owning the solution; or simply to get to the desired result faster. The architect needs to keep people engaged, elicit the necessary information, create shared understanding, and build consensus between all the stakeholders while keeping everybody focused. Decision-making is also the main part—the "bread and butter"—of the architect's job and the best way to come up with the right decision that will be supported by all participants is to use facilitation skills. Most of the time, the decision discussions would create or uncover conflicts that need to be labeled correctly and then the architect can facilitate a way out of them using a certain technique or framework.

The architect is also the one bringing consensus and finding common ground between various other practitioners as he or she is the one who best understands the big picture and how the solution will work for everybody. The "ivory tower" architect concept does not work well in today's world because developing a solution completely isolated from the team and then presenting it to them and telling them they should implement it would most definitely result in a failed project. A different and more successful approach is to come up with the solution as a group in a facilitated meeting. People will feel that they own the solution and are not just being told to implement it. This approach also increases trust and speeds execution of the entire project.

Thus, if facilitation skills are so important, how can we improve those skills? As with any other subject involving performance competencies and skills, there are many books that recommend certain methods and techniques. This section focuses on some of the most important points to help boost facilitation skills:

- *First and foremost, be prepared.* Take the time to understand who the participants are as well as what exact problem you are trying to solve. Be aware that being prepared does not mean trying to sway the conversation toward what might be the best solution. Trust is hard to gain and easy to lose, but coming across as jaded will most likely cause everybody to lose trust. The facilitator must be perceived as completely objective and not viewed as being in anyone's camp during a discussion.

- *Keep the participants focused on the problem.* Everyone has to agree with the problem definition by writing it in plain sight and referring to it whenever the discussion gets away from it. The facilitator must keep the discussion moving forward while at the same time not being so rigid that he or she frustrates meeting attendees. If the discussion drifts off to address a different problem statement or if the discussion becomes negative, bring it back on course.

- *Do the administrative work as well.* There are certain things to keep in mind during a session like establishing and maintaining a parking lot, encouraging participation, and maintaining a list of action items. This is an important part of facilitation.

The "parking lot" (the list of items parked for a later time) contains important issues that should be captured but are not germane to the problem you are trying to solve. Those need to be captured to be addressed in future discussions but put off to one side so the group can continue to focus on the problem at hand. Keeping other problems off to one side shows the team that their opinions are important. These items might also come up later in the discussion.

The facilitator also needs to encourage all attendees to participate in the session because some of them might not be eager to express their opinions (they might wait for others to speak, or simply because they are more on the "S" or "C" side; see Section 5.3). Although the facilitator needs to note the people who are not speaking, he or she must make sure to not bully them, which would create an environment of discomfort. Balance is key: sometimes others may speak too much; sometimes they may not speak enough.

Maintaining a list of action items is an important way to come up with the architectural solution because participation can be encouraged not just during the meeting but after it as well. Ensure that each action item addresses what needs to be done, who needs to do it, and when it needs to be done. Also, take the time to summarize action items at the end of the meeting to ensure everyone agrees as to their importance, assignment, and timing.

- *Ask the right questions and refrain from early decisions.* Facilitation is all about asking the right questions that are relevant to the problem. If the facilitator starts to lecture about a certain way to do things or the benefits of a certain option, he or she will most likely not get much out of the meeting as they will scare away others who have a different opinion or create early conflicts that will doom the proceedings. An architect will sometimes be seen as a leader—the "big guy or gal"— preaching what the workers must implement. A good way to eliminate this perception is to ask questions that guide the rest of the team to the solution instead of lecturing and telling others what to do. The architect should find a balance between asking questions and making decisions. Sometimes the architect continues to ask questions when it is actually the moment to jump in and make the decision. It is also about the way that the decision is made; stating things like "I am hearing that …" creates less conflict than something like: "Good, so we'll do … and you guys have to give me an estimate." Do not forget people tend to respond emotionally first.

- *Check rank or stripes at the door.* Although this expression comes from the military where they actually do have ranks and stripes, most companies have different levels of rank or responsibility as well. There might be practitioners (architects, analysts, and developers) but also management or executives all at the same meeting. The danger is that if the highest-rank participants express their opinion at the beginning, they subsequently set the course of the meeting to their agenda. Once they do that, those afraid to challenge them are not going to speak up. For everyone's sake, the facilitator must make sure that all the opinions have the same weight when it comes time to come up with the final decision and that no one opinion will override those of the other participants. Thus, it might make sense to talk in advance with the managers or executives who will participate in the meeting, letting them know (in a very polite way) about the implications should they would sway the conversation in a certain direction. One good way for them to participate to the meeting would be to stay for the first 10 to 15 minutes to set the stage and define the expectations and then to leave. Another good way for them to participate if they decide to stay on is to explicitly state that they are no more than regular participants and that their opinions should have no more weight than those of any other participants.

In the past, facilitation was viewed as a skill that was nice to have. Presently, it has become increasingly more clear that facilitation makes a big difference when we need to achieve results as teams. You do not need to become a professional facilitator to be a good architect. However, as with all other soft skills, paying attention to what is missing in your skill set is essential to your professional growth and development. Trying to constantly improve the way that you work with others and working hard translate into better work and more money.

References

[1]   Kepner-Tregoe Inc., *Executive Problem Analysis and Decision Making,* Kepner-Tregoe, 1973.

[2]   Ledington, P., *Soft Systems Methodology: Core Concepts,* PWJ & JN Ledington, 2014.

[3]   Inquisitive definition, http://www.dictionary.com/browse/inquisitive, accessed May 2016.

[4]   Grady, C. L., et al., "Neural Correlates of the Episodic Encoding of Pictures and Words," http://www.pnas.org/content/95/5/2703.abstract, accessed May 2016.

[5]   Onion, A., "Our Brains See Words as Pictures," http://news.discovery.com/human/life/our-brains-see-words-as-pictures-150324.htm, accessed May 2016.

[6]   Gouédard, S., "Leadership and Management Development: 7 Key Factors for Success," http://pmctraining.com/site/resources-2/leadership-and-management-development-seven-key-factors-for-success/, accessed May 2016.

[7]   DISC overview, https://www.discprofile.com/what-is-disc/overview/, accessed May 2016.

[8]   Communication definition, https://en.wikiquote.org/wiki/Communication, accessed May 2016.

[9]   Toastmasters International, https://www.toastmasters.org/, accessed May 2016.

[10]   Prosci, "ADKAR Model Change Management Model Overview," https://www.prosci.com/adkar/adkar-model, accessed May 2016.

[11]   Project Management Institute, http://www.pmi.org, accessed May 2016.

[12]   Slater, R., *29 Leadership Secrets from Jack Welch*, New York: McGraw-Hill Education, 2002.

[13]   Rothwell, W. J., *ASTD Models for Human Performance Improvement: Roles, Competencies, and Outputs*, Alexanderia, VA: ASTD Press, 1998.

[14]   Block, P., *Flawless Consulting: A Guide to Getting Your Expertise Used*, San Francisco, CA: Pfeiffer, 2011.

[15]   Mayer, J. D., M. A. Brackett, and P. Salovey, *Emotional Intelligence: Key Readings on the Mayer and Salovey Model*, Port Chester, NY: National Professional Resources/Dude, 2004.

6

How to Start: The Road to Becoming an Architect

Now that we have examined different kinds of architecture, and understand what a good architect should look like, it is time to discuss the path to becoming one. Although this book will not explore the various architecture frameworks and methodologies in great detail (because many other books already do a very good job covering these subjects), we will focus on the kind of knowledge needed to become an architect and the best way to achieve that goal. As stated in the introduction, this book is meant to provide the information you need to become a successful architect. It should also drastically shorten the time it takes to transition from another information technology (IT) profession into the role of the architect.

We will talk briefly about the different ways to pursue an architecture career, and then focus on architectural specific knowledge. Because there are various types of architecture (seen in Chapter 2), this information will vary according to each type. This book should be used as an executive summary to help identify what you need to focus on to become more proficient in the architecture domain in which you intend to specialize.

Under the general umbrella of IT architecture, you can find an array of jobs that differ considerably in their technical points of view. A complete description of the technical knowledge required for each of these would be impractical, owing to the diversity and depth required. We will settle for an overview and a summary instead, with lots of pointers to help explore those depths on your own.

6.1 The Many Ways to Become an Architect

The common technical knowledge that you need to become an architect often depends on what kind of architecture you are most inclined to pursue. This section will discuss each of the common ways to become an architect and what technical depth you would need to acquire along the way. The career path to IT architecture usually includes a combination of the roles discussed next. No matter your experience or the common technical knowledge required in your architectural domain, you can become a good architect by working on your architectural knowledge, as well as soft skills.

Unlike IT architecture's much older relative, construction architecture, IT architecture has only been on the market for a few decades. As such, IT architects do not always follow a clear career trajectory. For example, most architects have backgrounds as developers, infrastructure specialists, or business analysts (in the case of the business architect). Many architects do not even start out with a degree in IT architecture, simply because a junior architect first needs to get the technical and business knowledge that forms the foundation for an architecture career.

The main aspect that separates other roles from the architect is the level of detail. Figure 6.1 summarizes different kinds of architects and some of the roles from which these architects usually evolve (although there are other pathways as well.)

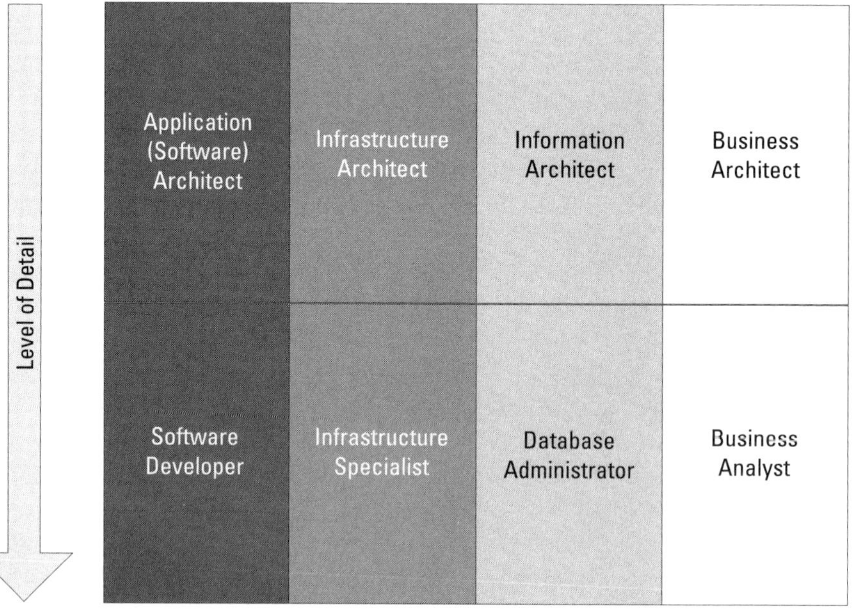

Figure 6.1 Separation of roles based on the level of detail.

Each of these architect roles will require you to build on your current background, and to expand your architectural breadth by learning about other technologies and systems. Recall the T-shape (Figure 4.1), which shows that becoming an architect (no matter what type) means sacrificing some depth in favor of breadth. You may have been a specialist in your domain (a software developer who specialized in developing Web services, for example, or a database administrator who knows how to fine-tune complicated relational databases). As an aspiring architect, you must expand your breadth and begin to explore other domains as well (such as infrastructure, database, and more). An application architect, for example, will have to elevate his or her thinking from an application model to the way that various applications and/or systems interact. Meanwhile, an infrastructure specialist will have to graduate from the level of Internet Protocol (IP) addresses and firewall configurations to develop an understanding of the network at a high level and the implications of using different zones of trust to separate the enterprise network.

Next we will discuss each of these paths to becoming an architect. It is important to understand that other paths might also be possible. For example, an infrastructure specialist could become an application architect or a security architect. There are no prerequisites or rules that can specify your path. You may need to work harder to gain the necessary background, but the same thinking applies to all types of architecture.

6.2　Architect with a Software Developer Background

One of the most common routes to becoming an application (software) architect is by first working as a software developer. This can help you to achieve depth in various programming languages and systems. The jump from senior developer to architect was once considered a natural evolution, because the architect was expected to know everything on the project from technical point of view so as to make the best decisions. Many developers may find a good fit in the architect role, because they have technical skills and knowledge and they prove themselves able to adapt their styles to hone important soft skills as well.

An architect should be the key decision-maker on a project, but this does not mean that this person needs to know anything and everything about the system. This should normally remain in the developer's domain (see the minimal and safe depth concepts in Chapter 4). An architect needs to act as the key decision maker from a technical point of view, and sometimes this occurs even if that person lacks the full body of technical knowledge in all relevant subject matters. As discussed in Chapter 2, an architect instead needs to rely on subject matter experts to get information, while using their soft skills to sell them on a proper solution, as well as the major stakeholders.

As an application architect, your work depends on your relationship with the developers and how well you understand their challenges and the obstacles they face. Although an architect should not be responsible for writing code, he or she should understand major design patterns and should be able to integrate them into their overall architectural solutions. Some would argue that if you are not able to check the developers' work, then what value do you bring to the table? However, being an architect is not just about checking the developers. There is a clear distinction in between a lead developer, who does code review and coordinates the work of the developers, and the architect, who should provide the overall solution including the interaction between the application components, what design patterns to use, implementation strategy, and so on. There may be extreme cases where the architect will fill in a gap and occupy the lead developer role as well, but these should be exceptions and not the rule. These are cases when you help the company and take ownership for the project as a whole to make delivery possible. As mentioned before, such exceptions might actually be the expectation for a small (maybe startup) company in which the need for separate job roles is not so acute.

Therefore, if you are a developer, what can you do to become an architect? In summary:

- Make sure you understand it and you like it. It cannot be overstated how many people decide to switch to an architect position, but then find out later that it was not what they thought or wanted. As with any career change, make sure you are familiar with your new role's expectations. As a successful developer, you hopefully have a good grasp of the role already.

- You need to do your apprenticeship and learn how things are done in the IT industry. This means working on "your depth" in one or more programming languages, modeling languages, business and industry knowledge, and systems. There are many things that experienced developers take for granted, but that you should also be exposed to such as the full software development life cycle (SDLC), various testing environments, how to promote the code, and for J2EE, for example, and how to deploy it on application servers. All of these are bread and butter for developers, but can be quite useful later on as an architect as well.

  Developing deep knowledge in programming languages can help you become a good hands-on developer, but that is not enough if you would like to become an architect (at some point). To possess sufficient architectural breadth later on, you must be exposed to modeling languages and be able to correlate the design with the code. Knowing a modeling language (and using it to document an application) can definitely help you to begin conceptualizing an architecture. Getting to know the other domains mentioned earlier is important, because you need to know more than just your application (how do the applications/systems interact?) and how the code serves the business. Whenever you normally interact with a business or systems analyst (or with business people directly), take this opportunity to find out more about what is behind the requirements, and how your new application (or application changes) could affect the business environment.

One should also understand what levels of detail to present based on the audience one faces. As a developer, you might not be required to present at L0, L1, or L2 (see Figure 6.2), but to become an architect you need to be able to understand all the levels of detail shown. Depending on how senior a developer is and how junior the architect is, there might be an overlap in between the two. Sometimes the developer may have a grasp on the application components and application services (not Web services), and the architect might have depth in this application, so that he or she might be able to contribute to design patterns, best practices, and so forth.

- Hone your soft skills. In addition to working on the technical side, one should also keep the soft skills presented in Chapter 5 in mind. Leadership, for example, is a skill you need as a developer (especially at the senior levels), as well as an architect. More leadership experience can put

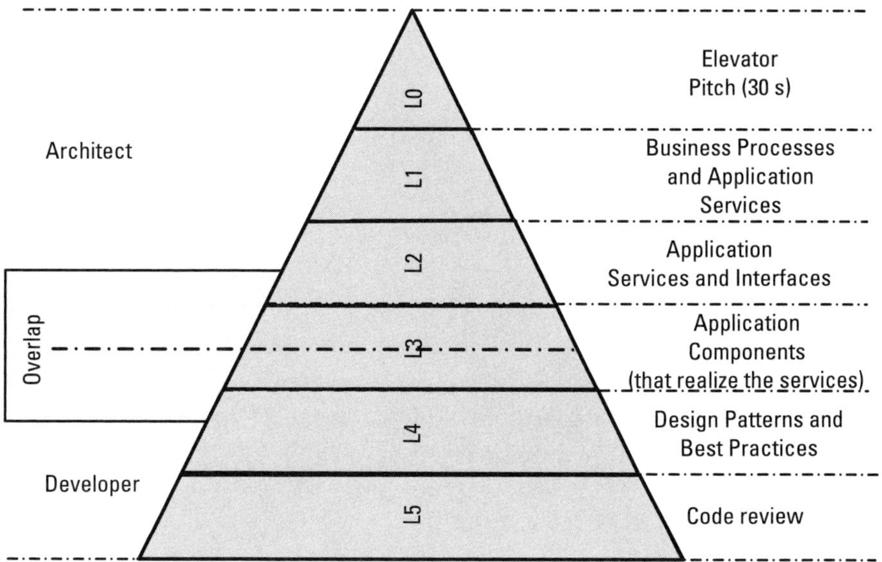

Figure 6.2　Level of detail for application architecture.

you in a good spot for promotion to senior developer, but only if you really know how to lead other developers.

You may not be the perfect communicator, negotiator, or facilitator at the beginning, but be aware of your personal development plan and work on getting better at each of the soft skills covered in Chapter 4. Take advantage of opportunities in which you must create a detailed design and sell it to the architect, project manager, analyst, and others to practice your craft. Use the soft skills mentioned in this book as a checklist and try to do a self-assessment or, better yet, get feedback from others you work with about your level of leadership, communication, presentation skills, and the remaining elements of the soft skills inventory.

- Give it a try. If possible at your company, try to request out-of-role assignments or take on the tasks of a junior architect. Talk with your manger and let him or her know that you are interested in doing architecture work. This is a sensitive step, because you do not want them to think that you are ready to leave the company tomorrow, but you do want to show your readiness to grow. Because many companies like to promote professionals from within their own ranks, with a little bit of luck you might get the assignment. Ideally, these assignments should expose you to the kind of deliverables that you will need to produce and generally what kind of work you will do as an architect. It may also be helpful to take a couple of soft skills courses like leadership, negotiation, and, most importantly, presentation skills.

- Being prepared strongly improves your chances of creating a good first impression as an architect and eventually making an out-of- role assignment into a permanent position. One of the reasons why companies are interested in promoting within the organization is because you have something they need. Over the years, you have likely accumulated quite a bit of business and industry knowledge, which puts you in front of other candidates from outside the company or industry. This is a win-win situation, where both the company and you have something to gain.

The other way to give it a try is not as risk-free. This may mean looking for your first architecture job outside of your current company. This is usually quite a big leap, because it is not easy to sell yourself if you have not actually practiced architecture at all. Ideally, you would have support from your current manager and work with an architect assigned to coach you. If this is not the case (assuming that you do not get any support from management), here are a few ways to bridge this gap:

- Ask to work with an architect from your current company and do a little bit of job shadowing to see how those theoretical concepts (like models, viewpoints, and more) apply to day-to-day work on the project.

- Pursue recognized architecture certifications that can help you understand and model different work scenarios. This is also a good way to prove to yourself (if you need it) that you are ready and that you will actually be able to do the work. The top five enterprise IT architect certifications, according to Tom's IT Pro [1], are based on the number of jobs in which the certifications were mentioned on well-known job sites (such as LinkedIn, indeed, and SimplyHired). They include:
  - Black Belt Certified Enterprise Architect (CEA) provided by the FEAC Institute.
  - Certified Service Oriented Architecture (SOA) Architect focuses on the design of the SOA technologies and the ability to integrate SOA solutions into general IT infrastructures.
  - CITA: IASA's Certified IT Architect provided by the International Association of all IT Architects, it focuses on the five pillars of architecture: business technology strategy, IT environment, design skills, human dynamics, and quality attributes.
  - Enterprise Architecture Center of Excellence (EACOE) requires candidates to demonstrate performance and give presentations in the classroom in the course of completing an EACOE Certification Workshop and associated modeling activities.
  - Open CA: Open Group Certified Architect provided by the Open Group, historically focused solely on IT and recently expanded the Open CA credential to cover business architecture and enterprise architecture as well.

- Last but not least, working for a small company where employees might wear multiple hats is another way to experience architecture work. Once you know what architecture really is, you may assess your role and realize that at least part of your job was actually application architecture, for instance. (This would apply mostly to senior developers.) Job descriptions may be less exact or accurate, even in medium to large companies. For example, a job such as technical specialist may have included some application architecture experience.

In summary, moving from the senior developer role to the role of architect should be a natural step up, as long as you understand what you are signing up for (no more coding), and you prepare for this step carefully.

6.3   Architect with an Infrastructure Specialist Background

As one of the many different types of IT architecture, infrastructure architecture deals primarily with the hardware part of information technology. Applications aim to automate business processes, but they also need to reside somewhere on various machines running different operating systems. Normally, infrastructure specialists or system administrators are those in charge of administering these systems. Similar to the software developer in the application architect track, the infrastructure architect can be considered a next step in the evolution from an infrastructure specialist. Architecture is the high-level view of all the systems, so in this case, the infrastructure architect would mostly deal with the major infrastructure components such as firewalls, storage devices, different infrastructure boxes, and so on. Ideally, an architect will have depth in at least one infrastructure domain such as networking, storage, Wintel, UNIX, and so forth.

As already mentioned, expanding your architectural breadth and using what you already know as a foundation is very important. Understand the difference between the architect's role and the specialist's role: It is no longer about the various configurations for some specific server, or for a network device, but about how all these things interact and work together to support access to the applications running on those servers. The purpose of the whole infrastructure is to serve the business applications that are running on it and, since the invention of the Internet, protect corporate and client information. The infrastructure architect's mandate is to design an infrastructure that provides all services required by the applications (and documented in the application architecture) in a completely secure and stable manner. As a specialist, you may have been primarily responsible for different pieces of the infrastructure, but now it is all about putting these pieces together and focusing on the high -level (infrastructure) design.

Although some advice from the previous section still applies, here is some additional guidance for an infrastructure specialist looking to take up the IT architect role:

- As with other roles, build on architectural depth and then focus on architectural breadth. Working as a specialist in one domain like networks or servers can expose you to the infrastructure world and how things get done in that world. Although most of this knowledge is specific to whatever technology/server/platform on which you are working, there are aspects that can prepare you for working on architecture, including how a certain network configuration could impact other servers and applications. As an example, blocking a firewall port might seem like a good method to tighten security, but if that port is being used by one of the company's Web applications, you just blocked your company from doing business over the Internet. Do not forget that, as an infrastructure specialist, you are the expert in your domain, and you know the most about your piece of infrastructure (and service that it provides). Trying to broaden your knowledge will benefit both the company (because you can vet the solution) and your career aspirations.

For an infrastructure specialist, working on breadth should also mean using tools from the application architecture realm, because the more familiar they become, the better you can design infrastructure to serve applications. This does not mean that you need to start learning a programming language, or do code reviews (you are not becoming a developer), but knowing enough about application programming interfaces (APIs) for each of the applications to see what kind of connectivity you need to provide to them, with whom they should talk, and what kind of requirements they might have from the infrastructure point of view (how big should the network pipeline be, what kind of infrastructure protocols are required, and so on) can be crucial.

- The gate to the architecture realm is soft skills. We will not go into this discussion in detail again, but these are on the list of mandatory skills to develop if you are thinking about becoming an architect.

- Also in the preceding section, preparing yourself by asking for an out-of-role assignment to practice your skills as an infrastructure architect can be helpful. This provides an opportunity to demonstrate that you can picture the whole company infrastructure, and you can exercise the strategic thinking expected of an architect. In such instances, try to start thinking at a high level before delving into the details of each of the infrastructure pieces.

- One of the most important factors that distinguishes the infrastructure architect from the specialist is the ability to consider the nonfunctional requirements when designing the infrastructure. Serving the application layer does not just mean complying with the functional requirements and providing a server that works, but also considering all the nonfunctional requirements that might have an impact on the way the infrastructure is designed. Some examples of the most important requirements of this kind include:

- Scalability: How can you design an infrastructure that is scalable enough for the current and future needs of the applications that will be deployed? As the architect, you have to imagine the best way to provide current and future scalability. For a certain machine, for example, it might be about deciding between going with vertical stability or going with horizontal scalability.

- Maintainability: As the architect, you need to provide a design that is easy to maintain. This might mean things like standardizing the operating systems and protocols that you will support.

- Performance: This is an important concern for you as the infrastructure architect, as you should ensure that the infrastructure will perform as per the service level agreement (SLA). When others are establishing the SLA based on the expectations that they have of the system, the infrastructure architect should be able to recommend something based on their previous experience. Do not forget that, just because the user does not want to specify what kind of performance they expect, it does not mean that they will not blame the architecture when the system is too slow.

- Reliability and recoverability: What is the user's expectation in terms of how quick the infrastructure component will be recovered in case of a disaster? The recovery time objective (RTO) is another important concern. Consider how your architecture solution will satisfy that requirement (by using either a failover mechanism or a cluster).

- Security: Many nonfunctional requirements can be grouped under security, such as auditability, privacy, and integrity. As with other nonfunctional requirements, you must first understand what is expected and then how your solution can satisfy it. It is important to take all security policies and standards into consideration when putting together an architectural solution. An infrastructure architect is responsible for security throughout all IT elements of the organization, and they must work closely with the information security office (ISO) that defines security policies and standards for the whole organization.

These are just a few of the important nonfunctional requirements to familiarize yourself with and to keep in mind whenever designing an architectural solution.

In summary, the main skill that can indicate that you may be a good fit as an architect is the ability to think at a higher level and see how all infrastructure components interact with each other.

6.4 Architect with a Systems Analyst Background

Another common path to the IT architect position is by evolving from a business or systems analyst role. The business architect role is sometimes considered to fall outside the IT department, but it is still worth discussing within the scope of this book.

The analyst's role can differ at each company; what is sometimes called a business analyst in some places is often called a systems analyst or a business systems analyst in others. For the purposes of this discussion, we will refer to the systems analyst role as the one responsible for knowing each system so that its occupiers can capture and document business requirements needed to implement a solution, and meet business needs.

Next, we focus on the path from a systems analyst to a business architect. Although much of this discussion can also be applied to the business analyst to business architect path, the main difference is the business analyst's lack of technical knowledge. Many of the competencies and techniques required of a business architect are the same as those needed for other (IT) types of architects, such as application and infrastructure.

As with the other career evolutions we have already discussed, the main difference between the systems analyst and the business architect is the level of detail (please refer to Figure 6.3). The role of the systems analyst is to document the current state of the business architecture, develop the target state based on the business goals and drivers, perform gap analysis, and define the roadmap to get from the current to the target state. How are these details translated from the technical world into the business world?

Although both roles focus on analyzing the business, the problem systems analysts need to solve is usually specific to a certain well-defined work package. Business architects define the work to be included in those work packages based on gaps between strategic needs of a business unit, and their abilities to meet those needs. The business architect, based on gap analysis, defines the business roadmap containing prioritized activities that are grouped into work packages, and that are later on grouped into projects. At an even higher level, this roadmap will have to be integrated with other roadmaps created for the application, data, and infrastructure, as the projects mentioned above will ultimately contain harmonized activities at the level of each domain architecture.

For example, the changes caused by creating a new business process will be packaged together with the changes needed to create the required application service and component. These changes are required to automate the new business process at the application level and with the new server (or logical partition) where the application will reside at the infrastructure level.

A systems analyst role usually requires more depth into the various lines of business, including business products, functions, and processes, while the

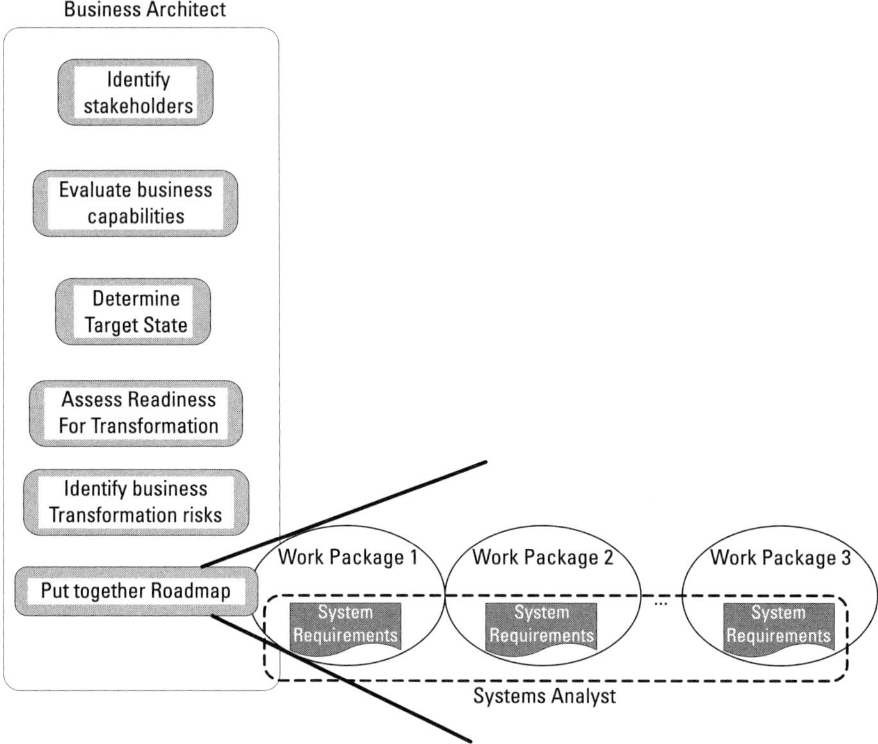

Figure 6.3　Business architect versus business analyst role.

business architect has a larger breadth. The business architect should be able to understand how adopting certain business goals and drivers would influence multiple lines of business.

One important distinction between the two roles is the strategic versus tactical thinking required. Business architecture is more of an ongoing process that happens in iterations, while analyzing future-looking strategies, capturing of capabilities, and modeling of inter-business and intra-business relationships, which are all needed to discover the key capability gaps that a business must be prepared to face. By comparison, systems analysts meet with business stakeholders and subject matter experts to determine business needs, to understand the systems, and to fulfill the IT requirements elicitation. As business and system subject matter experts, they should contribute to the definition of the current state and be able to take the strategic thinking produced by the business architect and apply it to day-to-day projects.

Now let us discuss what is expected from an employee looking to transition into the business architect role. We will only focus on advice that is specific to this path, and we will not delve into the common suggestions previously mentioned such as soft skills and finding a way to give it a try.

- *Get the background:* Similar to other domains, a business architect should know how to elicit requirements, as well as have a solid understanding of business analysis and business process modeling. You should be familiar with the main business modeling languages (BPMN, Testbed, or the Archimate business layer modeling, for example). This is needed to employ a common language to describe business services, capabilities, processes, functions, roles, and actors. Everyday duties for a systems analyst usually include use cases and use case modeling. Use cases are part of the Unified Modeling Language (UML) and will be discussed later in this chapter. In summary, it is essential to have a clear understanding of the concepts behind use cases and how can you use them to model requirements. Closely related topics include sequence and activity diagrams, used by systems analysts to describe flows for use cases. A business architect will not be required to put together these kinds of diagrams, but you need to have this background to understand the documentation that a systems analyst will put together to further detail your roadmap. Use case diagrams also provide a nice way of integrating with the IT world, as UML was initially intended to be used by software engineers, and there are tools that link the use cases and the flows with the application design. As a business architect, you may be less concerned with this, but this knowledge will come in handy in cases where you need to know more about the application layer that serves your business layer.

- *Become familiar with business architecture concepts:* No matter what framework or modeling language you use, you should be familiar with all the concepts and try to get some experience applying these concepts in day-to-day work. At a minimum, you need to know at least one of the architecture frameworks and modeling languages. Ideally, you would know a few in detail, be able to compare their concepts, and decide which one is the best for a given situation. You will also need to be able to:

  - Document the baseline description and analyze the current state often, from the bottom-up, also documenting the working assumptions about high-level architectures.

  - Develop an integrated view of the enterprise by selecting the correct viewpoint (business capability view, business knowledge view, business strategy view, business operational view) to demonstrate how stakeholder concerns are being addressed in the business architecture.

  - Determine what modeling language you will use to create the views identified in the previous step. For this step, and looking at the broader perspective and collaborating with the enterprise architects (if available), or the other architecture groups in the organization, try to standardize a modeling language that can be used to document all the domains of the enterprise architecture. An important benefit of this approach is the ability to ensure traceability among the business, application, and infrastructure layers.

- Develop target architecture to support business goals, drivers, and strategies. Identify the gap between the baseline and target states, as well as the new business processes, functions, services, and so on, as well as ones to eliminate.

- *Think strategically:* Business architecture, like most other architectures, is about strategic thinking. The business architecture is the basis for the subsequent architectures (application, infrastructure, and data). It is the foundation for all other architecture work. Thus, a poorly constructed business architecture would ultimately lower the success of your company. The real job of the business architect is to translate business strategies into a clear and solid business roadmap that can be further translated into the needs of other domains. Rather than thinking about the particular line of business and capability that you come from, aim to understand the needs of the whole organization. It is fair to say that both types of thinking (strategic and tactical) are needed, because you will use strategic thinking to design the roadmaps, and tactical thinking to follow up with those systems analysts who will implement those roadmaps within the systems requirements that they create for each project.

Entering business architecture from a more technical role requires a considerable expansion into business and industry knowledge (see Chapter 4), including common business capabilities, functions, processes and services, and the relationship of those functions, processes, and so forth (please refer to Figure 6.4). This might also mean becoming familiar with the business architecture concepts as specified above, as well as being exposed to business models and products common across the industry. For an insurance company, for example, understanding the various insurance products that are available, including life insurance and disability, is vital knowledge.

6.5 Architect with a Database Administrator Background

The term DBA, or Database Administrator, was first introduced by Michael Scott Morton, founder of a prominent MIT group, is defined as the person who centralized and controlled information of all kinds [2]. The role has since evolved, and nowadays it is mostly associated with a relational database like

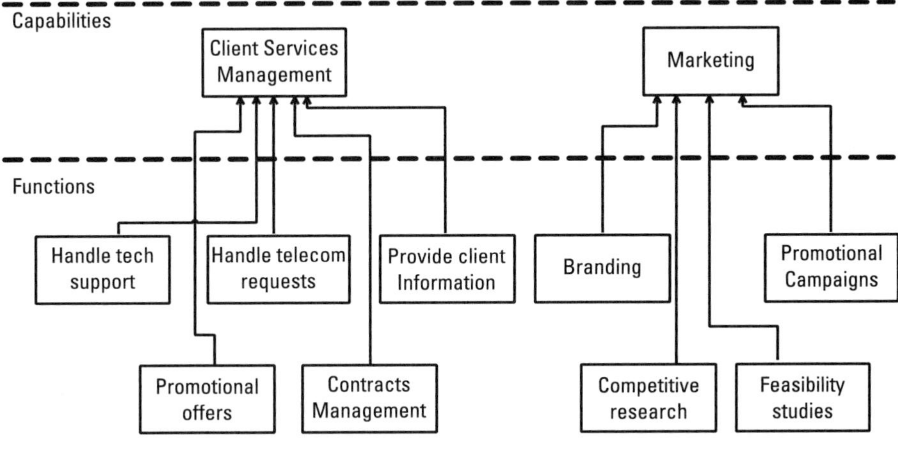

Figure 6.4 Documenting business capabilities and functions.

Oracle, SQL Server, and DB2. The duties expected of a DBA have also evolved into more than just the maintenance of the database. Now a DBA is also expected to suggest the best way to design the database so that queries against it run with maximum efficiency, as well as provide input into overall configuration of the database (for example, having a cluster, failover node, and other capabilities). To complicate things even more, there are also data analysts who analyze data from various sources and use specific data analysis methods, including data mining, text analytics, business intelligence, and data visualizations. Although this role is similar to that of the data architect, the data analyst is more focused on the analysis of the data, where the data architect focuses on designing the roadmap from the current state of data to the future target state.

Just like the name suggests, a DBA manages the administration of databases in terms of creating the instances, triggers, indexes, and other facilities. The most forward-thinking DBAs will ask themselves what they can do to proactively manage any performance issues they might have. Performance is one of the key, nonfunctional requirements, and perhaps the most important, because users get frustrated with long loading times, only to find out that the main culprit is the database.

The introduction of data warehouses and data marts have added more complexity to the DBA's role, because designing these things and as well as normal database administration can sometimes lead to overload. This creates a need for a data architect, especially in a medium-to-large company where there is enough work to be split among two roles. A data architect may know less about a certain product (Oracle, SQL Server, and so on) than a regular DBA, but they can focus primarily on designing databases, data warehouses, and other data manipulation tools or services. Even more so, the data architect knows not just how to design databases, but also how to do so in the context of multidimensional and object-oriented architectures.

Normally a data architect can complete data analysis, come up with a data model and then process a database design that satisfies both functional and nonfunctional requirements. They are not able to do this without the DBAs, who know the databases inside out. The DBAs would also be tasked with taking this logical data model then using it to build and maintain the database. A data architect is supposed to have the data breadth to come up with designs that provide a good fit for the overall enterprise architecture. Here again, it is not about what kind of work is more important for the company. Many data architects excel at both the administration and design, but the DBA's database knowledge allows them to perfectly fine-tune their databases to serve the application layer perfectly with the data it needs.

Now that the difference between the two roles has been established, let us explore some main points on the data architect checklist. As in the preceding section, we will not detail common items such as understanding the new job description, building soft skills, and using the opportunity to try it before changing jobs. The points discussed in detail in the following list items apply specifically to this career path.

- *Have a clear vision of the database in mind:* The data architect needs to have a clear vision of how the database should perform, how the logical design should be translated to a physical one, and the life cycle for data in the enterprise. They are the architects who know the most about how data is created, stored, used, and disposed of across the various types of databases. This should help them to establish data principles and patterns that must be followed for both logical and physical design. This clear vision includes using an established model for the data based on current and target state architecture, which can be referred to at any point in time. Creating such models requires familiarity with the various modeling languages for data architecture, as well as those spanning multiple domain architectures. This might mean going from describing data entities (or business data objects) and their relationships, to documentation of the logical model, and then reviewing and approving the physical model.

- *Be immersed in database design:* Over the years, a DBA will have acquired lots of database knowledge, which makes them the master of database administration. The next step up is to become familiar with overall database design. Learn how to get the information you need from a DBA, even if sometimes you might already know it. At the same time, focus on logical modeling, as well as data management processes. Data architects need to know how the SDLC applies at the data level in their company, because they will take part in various phases like data creation, storage, archiving, and purging. As a data architect, you will be involved less (if at all) in the database operations, because you will focus on the design and modeling efforts.

- *Understand interfaces to other domain architectures:* At the top end, a database design should serve the business processes/functions that the business or enterprise architect has designed previously. The data architect needs to understand these functions to properly model a data layer that can support those automated functions (please refer to Figure 6.5).

A similar relationship exists with the application architecture, because the data architecture supports application architecture by providing necessary data structures. This relationship also goes both ways: an application architecture also supports data architecture by providing applications to help populate the database, archive it, and purge it when no longer needed (data SDLC). Data architecture is also supported by the infrastructure architecture, which provides infrastructure services and devices for the database to live on (please refer to Figure 6.6).

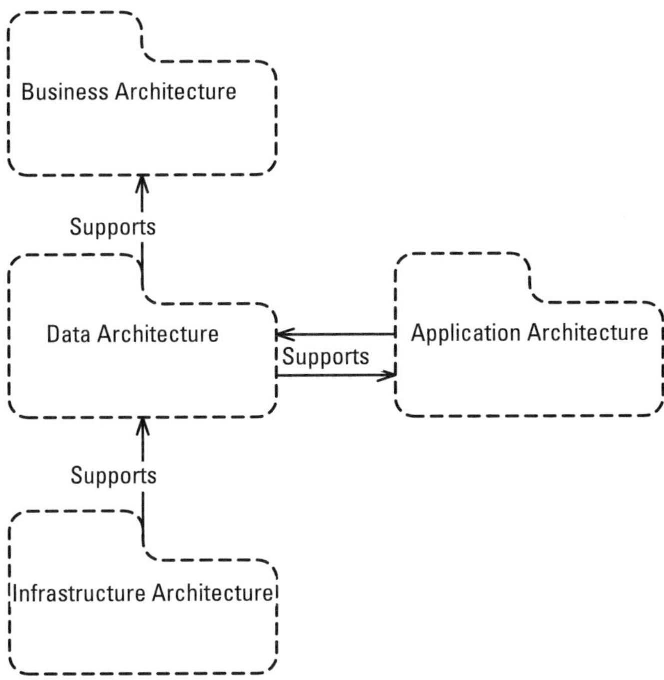

Figure 6.5 Relationships between the data and other domain architectures.

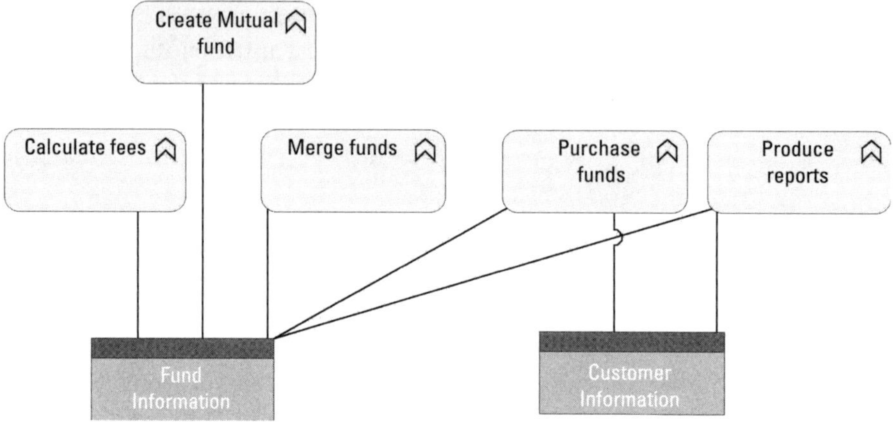

Figure 6.6　Example of business functions and data objects.

Collaboration with the business, application, and infrastructure architects is essential to define the support required, as well as how interfaces between the various architectures must work. In some cases, this might involve taking the overall vision created by an enterprise architect and refining it to come up with a database specific roadmap. This means documenting the current and target state of the data architecture, performing gap analysis, and defining the roadmap components specific to the data layer. This input can then define the enterprise roadmap, which combines the roadmaps from all other architecture domains.

- *Know the requirements and constraints:* Having a good grasp of the data roadmap also means defining the data interoperability requirements (for example, the XML schema for the XML file to be exchanged in between two application services). The data architecture layer will also impose certain constraints on the application and infrastructure architecture, and the data architect needs to document them, as well as their implications from both a functional and nonfunctional point of view. As the data architect, you own the database high-level design; therefore, you are also responsible for constraints imposed on other layers and what your design requires from other domains.

- *Always keep nonfunctional requirements in mind:* Architects should always ask themselves this fundamental question: How does the solution satisfy the nonfunctional requirements? For a data architect, this means:

  - Performance/scalability: Performance and scalability are some of the most important nonfunctional properties of a database. A DBA will be involved in tuning structured query langauge (SQL) queries to get the best performance, but the architect must be responsible for scaling the performance of very large databases and data warehouses to get the best connectivity. The best piece of advice is to rely on the DBA's expert knowledge and your data architecture breadth to make decisions such as determining the failover, deciding if a cluster is needed, and other key decisions. The answers to these questions, in combination with more fine-tuning (using the correct type of indexes and the correct type of transactions to reduce record locks) will help you design the best data solution for the specific problem.

- Extensibility: This means making sure information is readily available to the authorized users for decision-making, which is an important mandate of the data architect. Data storage should be extensible and able to adapt to an increase in volume or other new requirements.

  - Extensibility should be considered right from the design phase by creating a well-defined abstraction layer for client access and continue with the type of storage underneath the database layer. Adding a node to the cluster might be an option if you operate in a clustered environment.

- Security: Here is the key: not too much and not too little. This harkens back to the "boundaryless information flow" concept, first introduced by Jack Welch, which applies so well to data architecture [3]. This is all about getting the information to the right people, at the right time, in a secure manner. Most of the time, the same piece of data can be found in multiple places across the company. Your job as the data architect is not only to provide a good design to store it, but also to ensure the integrated information is protected(by combining multiple sources of information), as well as in collaborating with the security architect/officer to protect integrated access to that information.

The next sections will explore what kind of frameworks, modeling languages, and tools that data architects need to know, as well as topics on integration with other domain architectures.

6.6 Enterprise Architect

Enterprise architecture is all about maintaining balance with one foot in the business world and one foot in the IT world. This can be quite difficult, because these two distinct departments can move at different speeds and sometimes even in different directions. An enterprise architect is a step above a domain architect, because an enterprise architect needs a good understanding of application technology and technological concepts, as well as a good understanding of business architecture. Enterprise architects must also be able to think in broad business and IT terms, and be able to apply this knowledge to sell a vision of the enterprise and its benefits. Enterprise architects usually work as domain architects before moving into enterprise architecture. For example, a business architect can aim to gain the IT high-level knowledge needed to become an enterprise architect. Another option is to work as an architect under the IT umbrella and then, by becoming proficient in business knowledge, become an enterprise architect (please refer to Figure 6.7). Either way, it means moving from acting as a specialist in some specific architecture domain to becoming the architect for all domains in the enterprise.

In essence, this is a similar transition to the ones discussed in previous sections (from business analyst, software developer, database administrator, or infrastructure specialist to one of the domain architectures). The key to this transition is expanding your "architectural breadth" to include all the pieces of the enterprise puzzle.

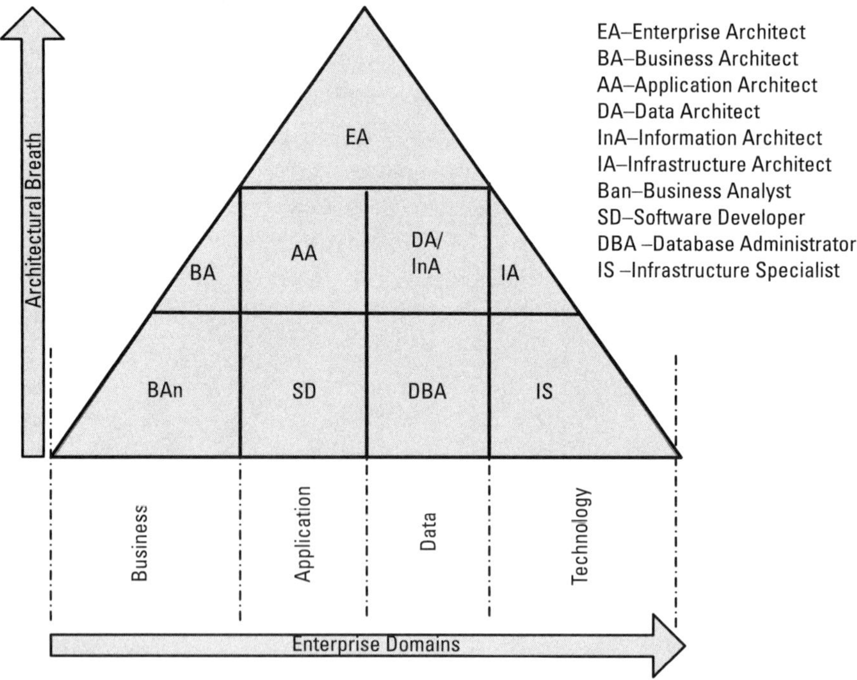

Figure 6.7	IT roles and the EA breadth.

The purpose of enterprise architecture is to provide a better integration between the business and IT, ensuring that each IT initiative has clear business value that will move the enterprise towards the target state. To achieve this, the enterprise architect must develop a holistic view of the enterprise that includes not only its IT services, but also the business functions and capabilities that they serve. An enterprise architect should know the application, data, and infrastructure components, as well as how they serve the business application layer. In this arena, the enterprise architect relies on the expertise provided by the other domain architects. There a few different paths to becoming an enterprise architect:

- This path assumes that you are an experienced business architect who already possesses a good understanding of business goals, drivers, constraints, and strategies, as well as how these are addressed by various business capabilities, functions, services, and roles. You may have the business piece of the puzzle wired, but you may still need to acquire IT high-level knowledge required to generate an enterprise solution that perfectly combines the two. More exposure to the application and technology world that serves the business layer is crucial for you. An enterprise architect needs at least minimal knowledge to be conversant in the application and technology language that specialized domain architects use in their architecture solutions. This hearkens back to the minimal depth concept that we discussed in Chapter 4, because you need to get at least minimal architectural depth in the various architecture domains so you can incorporate the high level of their architectures into your architectural view of the world.

- Start from any of the IT architecture domains. This assumes that you are already an architect who specializes in application, data, or technology domains. Becoming an enterprise architect is about expanding your architectural breadth to include the other IT domains, as well as business architecture. A key step in this transition is becoming proficient in the business language and learning to build business solutions that address business drivers, goals, and constraints. A good enterprise architect will begin thinking within the business domain and identify the various business components and then move on to creating the application, data, and technology architectures that fit the business.

No matter where you start, here are some common pointers to ponder:

- *Implement the business goals:* Your ultimate responsibility as the enterprise architect is to understand business goals and strategies, and be able to turn them into projects that implement the business vision. An enterprise architect is the orchestrator for any main changes within business IT, changes that can ultimately better position the company within the industry relative to its competitors. The work of the enterprise architect begins with identifying the next steps, getting into the other architecture domains, and seeing what kinds of application services and components are needed to automate business processes and functions and to support the application layer.

- *Become proficient in enterprise architectural frameworks and modeling languages:* It is essential to be familiar with frameworks and methodologies such as Zachman, TOGAF, and DODAF. We will examine these in more detail in Chapter 7, but the enterprise architect should know them inside and out, including their pros and cons and in which scenarios they best apply. Different companies use different frameworks (and some might have their own versions), but knowing these frameworks will allow you to master a common language. For companies that do not have a specific methodology or standardized deliverables, also may be possible to adapt standard deliverables proposed in one or more of these architecture methodologies. The enterprise architect's job is to tailor the enterprise architecture frameworks and methodologies to the specific needs of the enterprise. To make that happen, they should know or at least be familiar with most of them.

  The same notion applies to modeling languages. Here again, many companies use different languages for different kinds of domain architectures, so you need to understand the languages specific to your enterprise. Difficulties may arise when there might be a modeling language for business architecture, another one for data, and a totally different one for application and technology domains. As the enterprise architect, you need to understand all of them (at least at the architecture level, not necessarily at the detail design level) so that you can validate that all such information stays in line with your overall architecture.

- *Master the tools:* There are many tools to assist with implementing one or more modeling languages. Each such tool uses its own system to produce its models, views, diagrams, and so on. Some of these things can be quite helpful for documenting the current and target state for various domain architectures, as well as gaps, timeframes, and the work packages to address them. As with frameworks and modeling languages, there are quite a few tools available, so becoming familiar with some of them is essential. (More detail about these tools appears in Chapter 7.)

Apart from architect-specific tools, generic tools such as Microsoft Excel, PowerPoint, and Viso can help by producing charts, graphs, slides and diagrams that you can use to present and sell architecture solutions to major stakeholders.

- *Become a communication and relationship master:* As an enterprise architect, you will often deal with important stakeholders. Therefore you must possess or develop executive level communication skills. The chief information officer (CIO) is usually an enterprise architect's main point of contact. He or she will explain a problem (perhaps in vague terms), and the architect is the person who should know what questions to ask and what information is still needed to craft an appropriate architecture solution. Find out who will need to be convinced, and from whom you need to get buy-in. Such communication skills are crucial, allowing you to confidently interact with executives on a daily basis.

In summary, be aware that an enterprise architect is the only type of architect who spans the two very different halves of each company: business and IT. You must know them well and be able to confidently design for them not in isolation, but as two interrelated parts that affect each other. Without an enterprise architect, the business runs the risk of becoming overly complex, becoming expensive and inefficient. Without a flexible design that aligns both business and IT, a company can seriously injure its ability to respond to or anticipate market demands.

6.7 Other Types of Architects

Although most types of architects have already been mentioned, there are still others including systems, and security. This section summarizes guidance aimed at these types of architects.

6.7.1 Systems Architect

The responsibilities of a systems architect can encompass both the application and infrastructure domains, so either a developer or infrastructure specialist background can be helpful. Aim to expand your architectural breadth by working on initiatives that challenge you beyond your area of area of depth. As a systems architect, every solution will involve both application and infrastructure architectural changes to solve the problem. As a result, a systems architect might need to coordinate with an application or infrastructure architect to coordinate changes across these two architectural domains. This means that:

- Development background: Start building on your infrastructure knowledge to include infrastructure devices and services and not just applications and their interfaces. Explore how users are accessing your application and how is it distributed across multiple servers within the company.

- Infrastructure background: Seek out more information about the way applications work, how they are packaged, deployed, configured, and accessed by their users. In this way, you can build a complete picture about the information system that you want to describe without focusing too much on the details of how that application is built.

No matter what knowledge gap you are trying to fill, your objective is to present your architecture from different viewpoints based on stakeholders' concerns. You are not a subject matter expert; your job is to gather and present information that you have compiled, where you rely on those experts for accuracy. Let us use a car as a quick analogy: you might create a viewpoint showing its electrical system or transmission, but you are not the specialist who knows all the intricacies of these subsystems. You need to define the system from a structural and behavioral point of view, as well as how it fits into the existing environment.

6.7.2 Security Architect

In many medium-to-large companies, a security architect is often called the security officer. First, the scope of this discussion is limited to the role related to the IT domain. This role is part of the information technology department, but it is usually separate from standard project or enterprise architecture functions because of its nature. Although all architecture roles should be mindful of nonfunctional requirements related to security, security is primary main focus for the security architect. Such professionals are most focused upon:

- Creating, updating, and enforcing security policies and procedures: This should be part of the overall security architecture governance that is part of the compliance process. The first step in composing documentation that regulates security enforcement is to review all systems to gain an understanding of how they are used, who is using them, and where weak points may be found. Such policies and procedures are created to set a standard baseline for the way all security modules and checks are incorporated and followed by all architectures created for the enterprise. The security architect demonstrates the commitment of the organization to protect its data and to reduce (ideally eliminate) the threat of data theft or unauthorized access to key assets. When creating new policies and procedures, all applicable standards available for a specific location should carefully considered. An example might include a risk management policy, which protects the confidentiality, integrity, and availability of information and information assets, or the information access policy, which holds individual users accountable for their unauthorized or inappropriate access.

- Act as a security expert/consultant: A security architect may also be tasked with improving an outdated system or making recommendations to harden a relatively new system. These recommendations may include hardware and software upgrades, as well as new protocols for a system's users. As with other positions, a security architect should be able to communicate with other subject matter experts and gain the information they need to make good recommendations to other architects seeking to satisfy security requirements for their projects.

  Acting as a security expert implies consideration of all aspects of information security including ensuring the adequate protection of information assets in private and public networks, establishing formal security incident response tasks and steps, and ensuring that new products and services include the necessary security controls.

To become a security architect, architecture knowledge will need to be acquired, preferably from a specific training program. Such programs can explore not only the context of the IT world, but also the way architecture works, what the role expectations are, and how they can better interact with the other roles.

Starting out as a security expert, one has normally completed training on computer system security, or network security, gaining depth in one of the domains under the security umbrella. In such a role, a security expert will mostly focus on the electronic type of security: making sure a specific system, network, or device complies with specific security policies and procedures that apply.

However, as a security architect you are responsible for both physical and electronic security, though your focus may primarily be on the electronic side. A security architect specifies certain policies and procedures and then monitors their implementation through various projects. Expand your architecture security breadth to understanding the various system security, firewalls, and devices, as well as encryption algorithms and protocols.

Remember that, unfortunately, security architecture does not operate in a vacuum. Thus, there are a series of other constraints that can affect adoption of new security controls. For example, if the system is close to retirement and will be phased out in the near future, it may make more sense to waive implementation of a new security control. However, for a system with no retirement plans, it is a different story. A security architect needs capable soft skills to handle pushback: not having enough time or "it works the way it is" might be common excuses. Again, just being the system security or network security expert responsible for the solution is not enough. It also comes down to the ability to stand firm and to be the security watchdog when a tough stance is required.

Apart from security-related knowledge, a good understanding of the various architecture frameworks that exist is also required. Each framework has a separate module that deals with security. A security architect should be able to understand those modules and tailor them to their company's specific needs. Some frameworks also demand specific deliverables, or sections of a specific deliverable, that can be used. Such documentation is an important piece of the success of your security architecture and includes the way you communicate and reinforce your message. What constitutes too much or too little documentation is a common question. Consider what the minimal amount would before others get the architectural blueprints required to steer their architectural decisions, but not so much that they would be overwhelmed, or even worse, they might get contradictory information from different documents.

Risk assessment is particularly important to document when dealing with complex IT problems such as connectivity between various zones of trust and the outside world (the Internet). A security architect should be able to assess security risk and translate it into costs (losses for the company if risks actually occurred.) They should also provide architectural governance from a security point of view and make sure that each of the proposed architecture solutions documents and addresses security risks as well.

References

[1]   Best Enterprise Architect Certifications for 2016, http://www.tomsitpro.com/articles/enterprise-architect-certifications,2-640.html, , accessed May 2016.

[2]   Haigh, T., "A Veritable Bucket of Facts: Origins of the Data Base Management System," http://www.tomandmaria.com/Tom/Writing/VeritableBucketOfFactsSIGMOD,   accessed May 2016.

[3]   Ashkenas, R., et al., *The Boundaryless Organization: Breaking the Chains of Organizational Structure*, San Francisco, CA: Jossey-Bass, 2002.

7

Architecture-Specific Knowledge

In Chapter 4 and 5, we discussed about the various skills and competencies required as an information technology (IT) architect including technical competencies, business and industry knowledge, architecture design skills, and, last but not least, the numerous performance competencies needed to achieve the best outcomes working as an architect. Here we focus on acquiring architecture-specific knowledge including the architecture frameworks and methodologies specific to different types of architecture, architecture modeling languages, and so on. There we disclosed that, apart from common IT knowledge and soft skills, the core competency of any architect is an ability to produce solid architectural designs. This section focuses on the knowledge that such work requires, taking into consideration differences between different types of architects. For each of these types of knowledge, there are also certifications available. These will be mentioned only to observe which ones are relevant to specific kinds of architecture. Such pure architectural knowledge is necessary to allow you to build and document architectural designs at the application, infrastructure, business, or entire enterprise level.

Such knowledge comprises various architecture patterns, frameworks, and methodologies, as well as modeling languages that enable you to visually describe the models you create, and the tools that implement those languages.

7.1 Architecture Frameworks and Methodologies

7.1.1 Application Architecture Frameworks

As a result of the continually growing complexity of enterprise systems, and because system engineering has matured as a formal discipline, there is a clear need for an architectural view of systems. In the early days of this profession, this view was built using custom-made languages and frameworks, but soon it became clear that to move to a more capable maturity model and to institute repeatable processes, standardization was needed. This process began at the application level, because developers (mostly under the open source umbrella) started to leverage their initial custom-made frameworks to assemble the first application frameworks. The idea was not only to standardize, but also to increase the efficiency and speed to market of projects by reusing code, or even entire modules. Especially with the introduction of layering and complex Web applications, it became clear that a framework can be used to take care of such features as error handling, database connections, logging, transparent switching from one component to another, and asynchronous process execution. Using a framework at this level provides a way to capture best practices and patterns and ensure the compatibility of new technologies to existing components and services.

The introduction of frameworks can also be linked to object-oriented technologies, wherein an application is usually decomposed into application services and components. The major concepts of object-oriented technologies such as inheritance, composition, and interfaces are made possible by design frameworks that supply pieces of reusable design to control the flow between components and interfaces with the outside world. This eliminates the need to start from scratch, because code can be copied and pasted from previous applications. The architect is only supposed to provide the functionality that implements the requirements, and the architect should not have to worry about pieces of functionality common among all applications. This is a good example of a domain-specific architecture framework. Over time, more frameworks geared specifically towards specific domains have emerged. These frameworks and methodologies also incorporate some common patterns such as:

- *Layering:* In much the same way as machine architectures are built in layers, software systems also use the notion of layering first introduced in client-server systems. This was further refined by the introduction of a three-layer architecture. This consisted of a presentation layer, which was only concerned with interactions with the user, a domain logic layer, which performed calculations based on inputs and stored data and validated any data coming from the presentation layer, and finally, a data source layer, which was primarily responsible for storing persistent data. This structure then evolved into an N-tier, or multitiered architecture, which provides the flexibility of adding customized application or infrastructure layers as needed. Layering has been a popular pattern for many different technologies.

- *Service-oriented architecture (SOA):* SOA is an architectural model that aims to enhance the agility and cost-effectiveness of an enterprise, while reducing the burden of IT on the overall organization. The core concept of this model is a service. A service is the unit of solution logic that contains all capabilities or functions with a common purpose. Such a service has a defined exposed interface, and it is usually realized by an application component. The consumer of the respective service has no need to know about the way that the functionality is implemented, but only know about functionality that the service offers.

Apart from the way SOA is implemented, the concept itself should also influence the way a system is designed for all domain architectures. Specifically, with regard to technology architecture, implementing SOA can consist of a combination of new technologies (such as Web or representational state transfer (REST) services), as well as new platforms that specifically support the creation, execution, and evolution of service-oriented architecture solutions.

- *J2EE patterns:* With the introduction of Java technologies, a whole new set of patterns have emerged that concentrate on enterprise application architecture in the context of a layered architecture. These patterns can be individually used (refer to the Oracle Sun Java Center Web site, http://www.oracle.com/technetwork/articles/javaee/patterns-138526.html), or they can be combined to form larger solutions. They are the base for many various application frameworks that have emerged in the last 10 years.

- *Model View Controller (MVC):* This framework considers the model role. That role contains the data and behavior used by the user interface (UI), the view (a visual representation of the model), and the controller (which takes user input and manipulates the model to update the view appropriately). This pattern was a significant advance, because it provided a separation of concerns. As a result, it allowed for design of different presentations (rich client, Web browser, mobile device, and so on), along with the use of a consistent testing strategy.

  Apart from this well-known pattern, there are also a number of later patterns that can be classified as a Web presentation pattern, such as the front controller, template and transform view, and application controller.

As more application frameworks were developed, each of them implemented a subset of the patterns specified above from the different tiers. Every project has the same basic structure, represented by basic elements that need to be present: storing and retrieving data from a database, painting the user interface, security, logging, configuration support, monitoring, and automated testing. By using an extensible framework, preferably a vendor-neutral solution, you can cut project costs, improve software quality, and get a product to market faster. Although it is not within in scope of this book to discuss all the existing extensible frameworks, we will quickly touch on those which have played the most important roles in shaping application development over the last 20 years.

7.1.1.1 Java Technology Frameworks

The following are Java technology frameworks:

- *Struts 1:* Struts 1 has been available for nearly a decade, and was a huge success when first published. Struts 1 is an example of an MVC framework that implements its model in the form of Java Beans classes, its view as jsp pages, and the controller as an ActionServlet. Struts 1 uses a special Extensible Markup Language (XML) configuration file to store information about the various action classes and views that can be sent as the response to a Web browser. It uses patterns from the presentation tier such as: Front Controller, Dispatcher View, and View Helper.

- *Spring Framework:* The Spring Framework is an open-source application framework, and it works like an inversion of the control container on the Java platform. It is a lightweight framework for development of enterprise-ready applications. Spring can be used to configure declarative transaction management, or for remote access to logic using remote method invocation (RMI) or Web services, mailing facilities, and various options to preserve data within a database. The Spring Framework can be used in modular fashion; it allows usage in distinct parts, so that components which are not required by the application may be omitted. Spring is organized in seven modules:

  - Spring Aspect-Oriented Programming (AOP): One of the key components of Spring is the AOP framework. AOP is used in Spring to provide declarative enterprise services, especially as a replacement for enterprise Java beans (EJB) declarative services. It also allows users to implement custom aspects to complement their use of object-oriented programming (OOP) with object-oriented AOP.

  - Spring Object-Relational Mapping (ORM): The ORM package is related to the database access. It provides integration layers for popular object-relational mapping application programming interfaces (APIs), including Java data objects (JDO), Hibernate, and iBatis.

  - Spring Web: The Spring Web module is part of the web application development stack, which includes Spring model view controller (MVC).

- Spring DAO: The Data Access Object (DAO) support in Spring is primarily for standardizing data access work, using technologies like Java database connectivity (JDBC), Hibernate, or JDO.
- Spring Context: This package builds on the beans package to add support for message sources and for the Observer design pattern. It also supports the ability for application objects to obtain resources using a consistent API.
- Spring Web MVC: This provides MVC implementations for the web applications.
- Spring Core: This component provides Dependency Injection features. The Bean Factory provides a factory pattern that separates dependencies such as initialization, creation, and access to objects from actual program logic.

- *Hibernate:* This is an ORM tool that maps domain objects to a relational database, creating a layer between the database and the application. To preserve the data, Hibernate creates an entity class. It uses configuration files that connect to the database information, class mappings to tables, and so on. Hibernate also provides built-in transaction APIs, which abstract the application away from the underlying JDBC or JTA transactions. Each transaction represents a single atomic unit of work. One session can span multiple transactions.

- *Grails:* Grails is an Open Source, full stack, Web application framework for the Java platform. It takes advantage of the Groovy programming language and convention over configuration paradigm (design paradigms used by frameworks to decrease the number of decisions that a developer is required to make without necessarily losing flexibility; a developer only needs to specify aspects of the application) to provide a productive and streamlined development experience. Grails contains the following modules:
  - GORM: An easy-to-use Object Relational Mapping layer built on Hibernate;
  - GSP: An expressive view technology called Groovy Server Pages;
  - A controller layer built on Spring MVC;
  - Gant: A command line scripting environment built on Groovy;
  - An embedded Tomcat container configured for on-the-fly reloading;
  - Dependency injection with a built-in Spring container;
  - Support for internationalization (i18n) built on Spring's core MessageSource concept;
  - A transactional service layer built on Spring's transaction abstraction.

Grails defines the notion of a service layer. Grails best practices discourage embedding of core application logic inside controllers, because those neither promote reuse nor a clean separation of concerns. Services in Grails are the place to put the majority of the logic in an application, leaving controllers responsible for handling request flow with redirects. A service contains business logic that can be reused within any Grails application.

These frameworks must be understood in detail by developers, but an application architect will need to know their overall architecture and the main patterns that each respective framework uses to create an application architecture.

7.1.1.2 Microsoft .NET Technology Frameworks

Originally introduced as a single framework, the .NET framework subsequently split into .NET compact frameworks with a smaller footprint that could be used for smaller devices, such as Windows Phone. Since then, .NET frameworks have split even further into different subframeworks that include Windows Store, Silverlight, ASP .NET4, and ASP .NET5. The core principle behind having multiple frameworks is so that they can be deployed as a single unit, meaning that factoring is no concern. .NET frameworks used to be composed of different implementations, with nothing shared between them. In its most recent incarnation, .NET provides a well-factored implementation that now forms the core of the current .NET framework.

The main component is the .NET Core, available in Open Source form and supported by a variety of operating systems, including Linux and Mac OS X. It has an optimized implementation composed of a unified Base Class Library (BCL) that contains standard programming features such as Collections, XML, and DataType definitions. At the bottom of the .NET technology stack, we find a Runtime Adaptation Layer, a .NET Native Runtime (specific to the .NET environment), and CoreCLR used by ASP.NET5. At the top of the technology stack, we find application-specific APIs, such as the Windows Store App model and ASP.NET 5 App model. .NET Core is designed to be modular, and as such, it forms the foundation for any .NET verticals.

.NET also has design patterns specific to different programming languages. Some useful reference books include *.NET Design Pattern Framework 4.5, .Net Framework 4.5 Expert Programming Cookbook*, by A. P. Rajshekhar, or *Professional ASP.NET Design Patterns,* by Scott Millett. Apart from the three groups of design patterns, *.NET Design Pattern Framework 4.5* contains other specialized patterns such as Active Record, Unit of work, Head First, and Multi-tier, as well as the patterns mentioned before in the Java world, such as Model View Controller (MVC).

Gang of Four (GoF) design patterns are considered the foundation for .NET patterns. *Design Patterns: Elements of Reusable Object-Oriented Software*, written by Erich Gamma, Richard Helm, Ralph Johnson, and John Vlissides, is a classic reference that describes various development techniques and pitfalls, in addition to providing 23 object-oriented programming design patterns [1]. It has laid the foundation for many design patterns in different programming languages, which may be divided into three categories:

- *Creational patterns* provide ways to instantiate single objects or groups of related objects. They include Abstract Factory, Builder, Factory Method, Prototype, and Singleton.

- *Structural patterns* provide a manner to define relationships between classes or objects. They include Adapter, Bridge, Composite, Decorator, Facade, Flyweight, and Proxy.

- *Behavioral patterns* define manners of communication between classes and objects. They include: Chain of Responsibilities, Command, Interpreter, Iterator, Mediator, Memento, Observer, State, Strategy, Template Method, and Visitor.

A .NET application architect not only has to know the .NET framework in depth, but he or she must also know the design principles and patterns for developing successful solutions on the Microsoft platform, design a solution's layers, components, and services, and also select the key quality/engineering attributes.

7.1.2 Enterprise Architecture Frameworks and Methodologies

An enterprise architecture (EA) framework defines how to create and organize views and objects associated with some specific enterprise architecture. An architecture framework provides the guiding principles used to establish common practices for creating, interpreting, analyzing, and using architecture descriptions within a particular domain and/or layer. An enterprise architecture framework structures the architect's way of thinking in some specific area with supporting catalogs, diagrams, matrices, and models. One of the key benefits of enterprise architecture is its ability to support decision-making in changing business situations. Because enterprise architecture brings together business architecture and IT architecture within a single cohesive model, it is possible to trace the impact of changing business goals, drivers, and strategies, and also the impact of such changes on the systems impacted thereby.

Early enterprise architecture frameworks included the Zachman framework in 1987, as well as TOGAF and its early predecessor, TAFIM (each of these will be explored later). Some of these frameworks also contain a methodology for applying that framework and deliverables to be produced (including templates for each of these). Although this chapter does not detail an extensive comparison of these various frameworks and methodologies, it should provide a good starting point to learn more about them.

The ISO/IEC/IEEE 42010:2011 standard (see Figure 4.5), which focuses on information systems, provided the first way to standardize the various objects needed to describe the enterprise architecture and its relationships. Based on this standard, existing frameworks and methodologies were realigned to accommodate integrating various domain architectures. It is common to recognize three or four types of architecture across all such frameworks, each corresponding to a particular domain including:

- Business architecture;

- Information systems architecture, often subdivided into data architecture and application architecture;

- Technical architecture.

Beginning in 1996, a number of federal laws were passed in the United States to improve the way that the federal government acquires, uses, and disposes information technology. These laws were made largely in response to a number of major corporate and accounting scandals that cost investors billions of dollars when share prices of affected companies fell. The major acts of law that shaped enterprise architecture frameworks include:

- *Clinger-Cohen Act:* Requires each agency head to establish clear accountability for IT management activities by appointing an agency chief information officer (CIO) with the visibility and management responsibility necessary to carry out the specific provisions of the act. The CIO should drive reforms to:
    - Help control system development risks;
    - Better manage technology spending;
    - Succeed in achieving real, measurable improvements in agency performance.
- The Sarbanes-Oxley act requires that a government information technology shop operate just as an efficient and profitable business would operate. It emphasizes an integrated framework of technology aimed at efficiently serving the business [2]. The Sarbanes-Oxley Act assigns personal responsibility to senior management of public and nonpublic organizations in the United States and other countries. IT organizations

that do not implement effective architecture governance will be unable to achieve a high level of architecture compliance, and they will have no effective means to manage to business goals, or to attain a meaningful degree of cost control [3].

Although the first enterprise architecture framework (Zachman) was published almost 30 years ago, the introduction of these acts suddenly precipitated the introduction of new enterprise architecture frameworks, or revisions of the old ones, that would address governance concerns. Although most of these frameworks were introduced to address the need for more manageable systems, systems that would not hinder an organization's ability to respond to current and future market conditions in a timely manner, some of the modules can also be applied to any of the four domain architectures individually.

For example, the application architecture phase from TOGAF can be used by application architects to document the current/target state of the application/system, as well as the gaps and the roadmap. It contains the steps to follow, the inputs needed, and also templates for the documentation to be produced. Even if enterprise architecture is not your aspiration, a general understanding of at least one of these frameworks, and how they apply to your specific domain, is essential.

7.1.3 Frameworks Most Commonly Used

This section focuses on major frameworks currently in use at many companies. Please refer to the bibliography at the end of the book for more information about each of these.

- *Zachman framework:* This was first introduced by John Zachman as a framework for information systems architecture. It is a logical structure for classifying and organizing descriptive representations of an enterprise, which can be significant to executives and information systems developers. Its layout is simple: a two-dimensional matrix with questions on the horizontal axis (what, where, when, why, who, and how) and levels of detail on the vertical axis (contextual, conceptual, logical, physical, detailed).

  This framework's purpose is to translate abstract ideas at the scope level into concrete instantiations. The framework summarizes a collection of perspectives involved in enterprise architecture by representing them in the matrix. Its rows help to define the types of stakeholders, while its columns identify various aspects of the architecture.

The Zachman framework is not a methodology, as it does not imply any specific method for collecting, managing, or using the information that it describes. It is easy to understand, it addresses the enterprise as a whole, it is defined independently of tools or methodologies, and any issues can be mapped against it to understand where they fit. One important drawback is the large number of cells, which can be an obstacle to the framework's practical applicability. Also, relationships between different cells are not well specified (see Figure 7.1).

- *The Open Group Architecture Framework (TOGAF):* This is a framework, a detailed method, and a set of supporting tools all in one package. It is meant for architects developing an enterprise architecture. The original development of TOGAF Version 1 in 1995 was based on the Technical Architecture Framework for Information Management (TAFIM), developed by the U.S. Department of Defense (DoD) [4]. The DoD gave The Open Group explicit permission and encouragement to create TOGAF by building on the TAFIM, which itself was the result of many years of development effort.

   TOGAF is composed of seven main modules that provide:

   - A high-level introduction to the key concepts of enterprise architecture, including the definitions used throughout TOGAF (see Figure 7.2).

   - Architecture development method (ADM): This is the methodology included within TOGAF. It is a step-by-step approach to developing an enterprise architecture, but certain subchapters can be used to create architecture for the subdomains (like business, application, data, technology). It provides guidelines and techniques for applying the ADM.

   - Content framework: This includes a meta-model for architectural artifacts, use of reusable building blocks, and templates for the main architectural deliverables.

   - Enterprise Continuum: This discusses categorizing and storing outputs from the architectural activity. It includes discussion on how to organize an architectural repository.

   - Reference models: These include two out-of-the box architectural reference models (a technical reference model and an integrated information infrastructure reference model).

   - Architecture capability framework: This module is a great help when setting up an enterprise architecture function within your organiza-

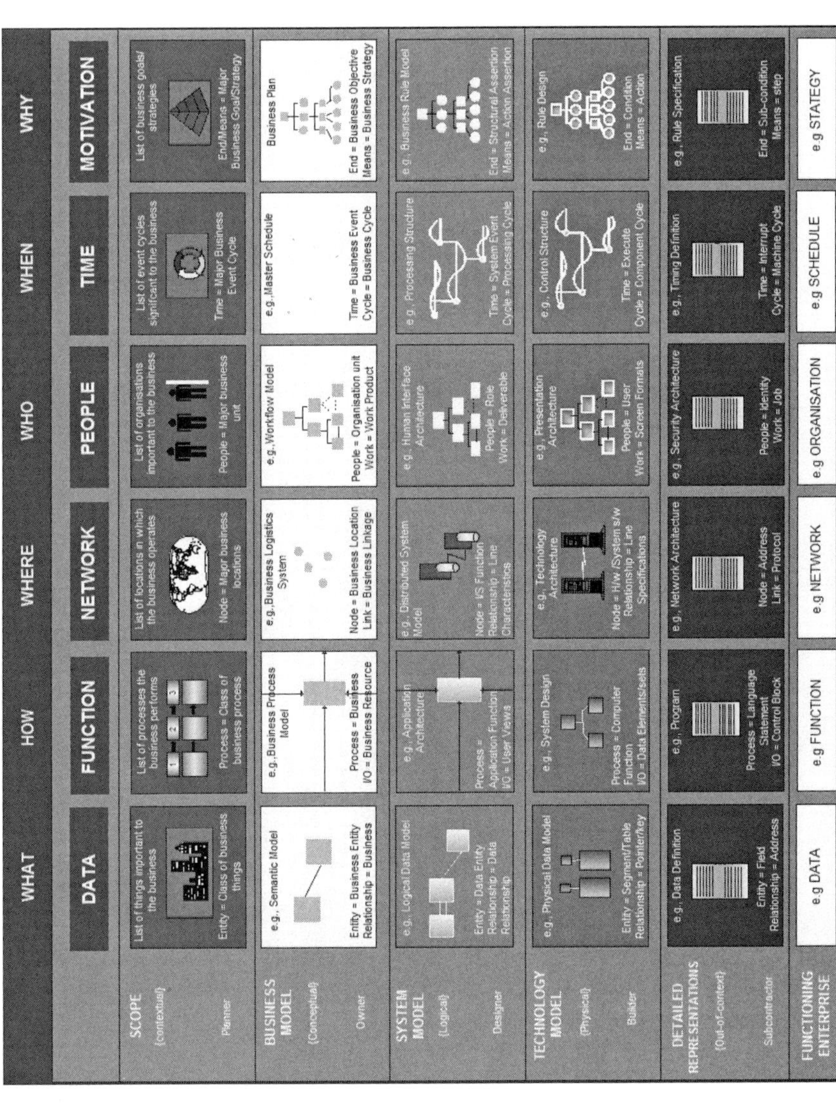

Figure 7.1 Zachman Enterprise Framework. (Published with the permission of John A. Zachman and Zachman International®, Inc., www.zachman.com.)

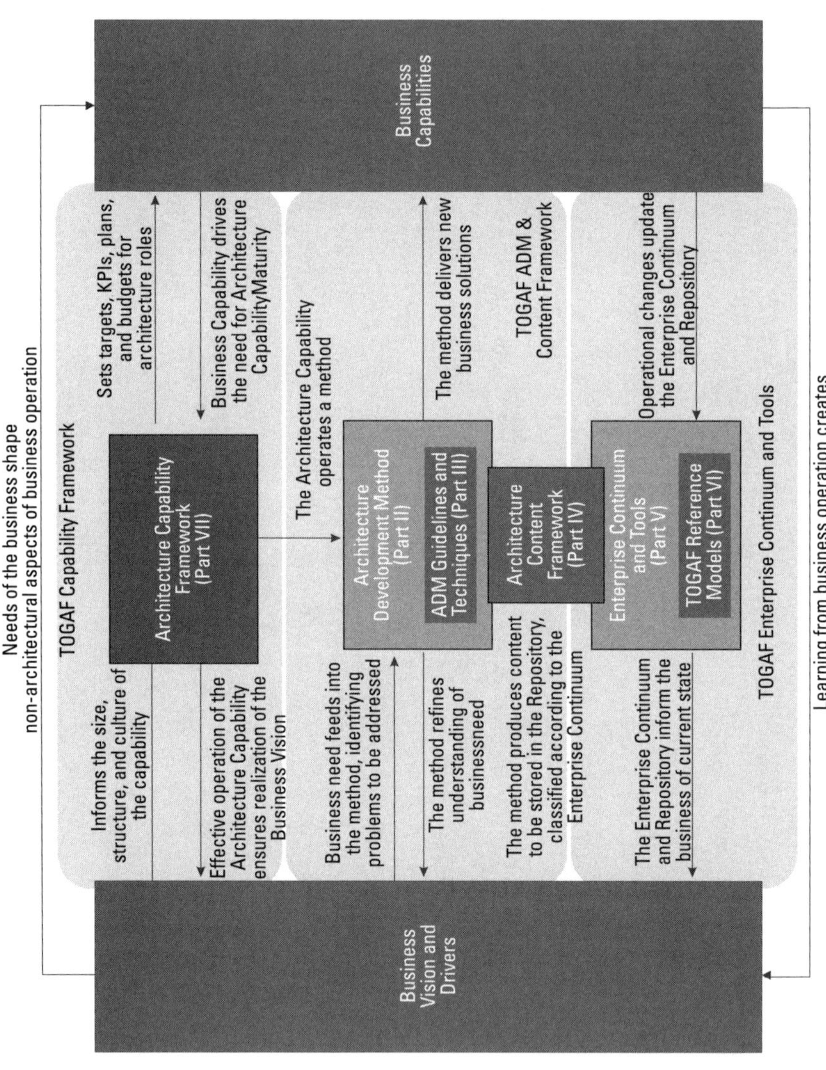

Figure 7.2　Structure of the TOGAF framework. (Source: TOGAF® version 9.)

tion, because it discusses the structure, processes, skills, roles, and responsibilities required to establish and operate it.

TOGAF is designed to support the four architecture domains commonly accepted as subsets of overall enterprise architecture: business, data, application, and technology architecture.

The core module of TOGAF is the ADM (please refer to Figure 7.3) that provides a tested and repeatable process for developing architectures. All of these activities are carried out in an iterative cycle of continuous architecture definition and realization and allow organizations to transform their enterprises in a controlled manner in response to business goals and opportunities.

The most important thing about TOGAF is that it can be tailored to the needs of your organization. For example, TOGAF can be used in these ways:

- Terminology tailoring: Use the terminology that is already understood across the organization. This means that sometimes TOGAF terminology will need to be translated into similar but familiar terms.

- Methodology/process tailoring: TOGAF provides a generic methodology. The first task should be to tailor the processes and align it with the other processes and methodologies that already exist like project and portfolio management, operations handover, and procurement processes.

- Content tailoring: The content of the framework can be customized for the specific requirements of the enterprise.

Imagine TOGAF as more of a Swiss Army knife than a cookie cutter, that is, you need pick only the content that is important for your domain architecture. Whether you need business, application, or infrastructure architecture tools, there are steps you can follow to put together the current/target state, as well as define the gap and roadmap. Once these concerns are outlined, there are lots of viewpoints to choose from to help you show how the business, application, data, or infrastructure architecture satisfies those concerns. During the implementation phase, there are also other tools to ensure proper governance and to make sure that the right vision is what actually gets implemented.

- *OMG's Model-Driven Architecture (MDA):* MDA aims to provide a vendor-neutral approach to interoperability. It is built on existing object management standards such as Unified Modeling Language (UML) and the meta object facility. MDA serves to raise the level of abstraction at which software solutions are specified. This is done by defining a

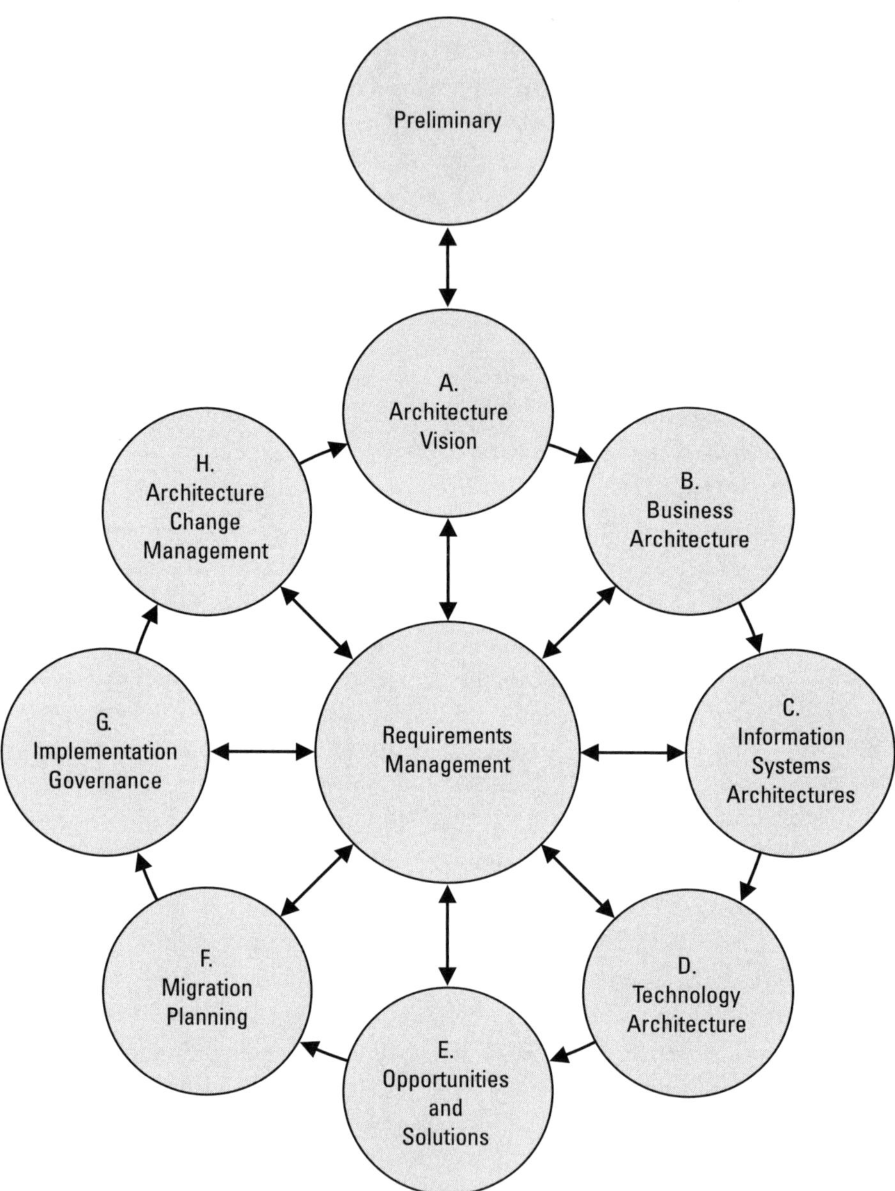

Figure 7.3　TOGAF architecture development method. (Source: TOGAF version 9.)

framework supported by a collection of standards, which, in turn, sets a new benchmark for generating code from models, and vice versa. MDA-based software development tools already support the specification of software in UML, instead of in a programming language, such as Java.

Table 7.1
Phases Within ADM

Preliminary Phase	Includes preparations required to meet the business directive for a new enterprise architecture, including the definition of an organization-specific architecture framework and the definition of principles.
Phase A: Architecture Vision	The initial phase of an architecture development cycle. It includes defining the scope, identifying stakeholders, obtaining approvals, and creating the architecture vision.
Phase B: Business	Business architecture is developed to support the agreed architecture vision.
Phase C: Information System	Develop information system architectures, including data and application architectures.
Phase D: Technology	Develop technology architectures.
Phase E: Opportunities and Solutions	Conduct initial implementation planning and identify delivery vehicles for the architecture previously developed.
Phase F: Migration Planning	Formulate detailed sequences of transition architectures with a supporting implementation and migration plan.
Phase G: Implementation Change Governance	Provide architectural oversight to the implementation.
Phase H: Architecture Change Management	Establish procedure for managing change to the new architecture.

The focus of this framework has been expanded to include business aspects of an enterprise. MDA now comprises of three abstraction levels with mappings between them:

- The requirements for the system are modeled in a computation-independent model (CIM), which describes the situation in which the system will be used. Such a model is sometimes called a domain model or a business model. It hides most or all information about use of automated data processing systems.

- The platform-independent model (PIM) describes the operation of a system, while hiding details specific to any particular platform. A PIM shows that part of the complete specification that does not change from one platform to another.

- A platform-specific model (PSM) combines the specifications in the PIM with the details that specify how that system uses a particular type of platform.

MDA has already published its own language for business processes, and UML will be used for the platform-independent and platform-specific models.

The key feature here is the notion of mapping as a set of rules and techniques, which can be used to modify one model to get another model. In certain restricted situations, a fully automatic transformation may be possible.

- *Federal enterprise architecture (FEA):* FEA has a long tradition behind it. It is an attempt to unite various government functions under a single, common enterprise architecture. It is quite comprehensive and has both a broad taxonomy and methodology (like TOGAF). In fact, it can be viewed as a methodology for creating enterprise architecture that can be also applied to private companies.

FEA consists of five reference models: business, service, data, components, and technical:

- Business reference model: This provides a business view of the several functions of the enterprise.
- Technical reference model: This defines the various technologies and standards that can be used to build the IT system.
- Components reference model: This provides an IT view of the systems that can support business functionality.
- Data reference model: This provides the standard way of describing the data.
- Performance reference model: This is a standard way of measuring the business value delivered by the enterprise architecture.

FEA also has a few other modules:

- Segment model: This gives perspective on how the enterprise architectures should be built.
- Creation of enterprise architectures: This includes architecture analysis (that maps to the Architecture Vision phase in TOGAF), architecture definition (definition of the technology architectures), investment and fund strategy (how the project will be funded), and program management planning (including the performance measurements).
- A transitional process guide to migrate from pre-enterprise architecture to post-enterprise architecture.
- Taxonomy for cataloging assets.
- Key performance indicators (KPIs) used to measure the success of the enterprise architecture and drive its business value. KPIs are mostly concerned with architectural maturity, architecture use, and architectural results. In terms of architecture maturity, levels are mapped to the capability maturity model integration (CMMI) levels (initial, managed, defined, quantitatively managed, and optimizing) [5].

The two core concepts of FEA are:

- An enterprise is built from segments, which are major lines of business functionality. There are two kinds of segments: core mission area segments, which are vital for the purpose of the particular area within the enterprise, and the business services segment, which is foundational to most types of enterprise (financial management, for example).

- Enterprise services represent well-defined functionality that might span multiple lines of business (for example, security management). They are not the same as services in the service-oriented architecture (SOA) model, because they are much broader and span more than just a technical implementation.

FEA complements the other frameworks described, because it includes methods for measuring business benefits, as well as detailing how architecture can be adopted across an enterprise.

- *Gartner:* Gartner is more of a methodology, and it does not include taxonomy (as does the Zachman framework, for example). Gartner methodology is a practice to be implemented across an organization with the help of Gartner enterprise architecture consultants. This framework operates according to the following principles:

  - Enterprise architecture is an ongoing process and not just a few artifacts.

  - Designing enterprise architecture is possible only by bringing together business owners, information specialists, and technology implementers, and giving them a common vision that drives business value. The success of a design can only be measured in pragmatic terms, such as increased profitability.

  - Any architecture effort should begin with defining the business strategy and its drivers. Strategy should be at the base of any enterprise architecture efforts and not just the engineering of the solution.

  - The architecture should have support from the highest levels (executives) of the enterprise; otherwise, it will not succeed.

  - As with Zachman, Gartner provide no guidance on implementation processes. Rather, it only serves to help with the enterprise architecture design phase.

  - Gartner provide no process to establish enterprise architecture practice within the organization.

Since 2012, Gartner has been working on revamping its architecture framework by introducing business outcome-driven enterprise architecture, which defines a simple and easy way to deliver enterprise architecture in three phases [6]:

- Target: Identify the target business outcomes and the outcome-driven success measures.
- Frame: Tailor your enterprise architecture framework and link information chain dependencies.
- Plan: Define the enterprise architecture process and identify the task interdependencies.

Here, a major guiding concept is disruption, which can be social, political, economic, and environmental. Disruption challenges the fundamentals in any organization. One would rightfully ask: how can an architect prepare an organization and proactively design to meet such challenges? The answer lies in the fundamentals of enterprise and application design: be adaptable (collaborative and change-oriented), keep simplicity in mind (focus on the modular design, standard processes and interfaces, and lightweight governance), and focus on visibility (proactively seek and analyze information about the ecosystem and proactively track and analyze risks).

Gartner's model suggests that during the process of transforming an enterprise to match the target, there will most likely be business disruptions. As an enterprise architect, you must prepare the company to be able to assess and scale the impact of such a disruption. The architect develops an organization's ability to survive and adapt. Usually, disruptions and uncertainties pull an organization away from the desired business direction, just as new regulatory issues or disruptive technologies (tablets and smartphones) can impede, or slow, new product development and the desire to increase operating efficiency.

In summary, the Gartner framework focuses on showing business value by delivering real outcomes or signature-ready recommendations and, most importantly, positioning enterprise architecture in terms of investment. Currently, the only way to adopt and implement Gartner is by relying on Gartner consultants who have the expertise to deal with such large enterprise architecture transformation exercises.

- *Rational Unified Process (RUP):* RUP is more of an iterative software development process. All of its elements can be tailored to fit a company's needs.

This framework is based on a set of building blocks. These building blocks describe what will be produced as part of software development, what skills are required, and a step-by-step description of how specific development goals are to be achieved. RUP uses four project life-cycle phases. Each phase outlines one key objective, or milestone, which should be accomplished to move onto the next phase. Those phases are:

- *Inception:* The key objective of this phase is to scope the system effectively so as to validate cost and budget. This is the phase where a business case is established, including a basic use case model, risk assessment, project description, and success factors. The most important criteria to meet include a good understanding of the requirements, the depth and breadth of the architecture, and also establishing a project baseline.
- *Elaboration:* The key objective is to mitigate primary risk items identified in the inception phase. The outcome consists of a use case model, description of the application architecture (including technical risks and their mitigation and eventually using prototypes to demonstrate this), a revised business case model, and also a development plan.
- *Construction:* The key objective is to build the system and produce a first release of the application.
- *Transition:* The key objective is to transition the application to production support, train users, and check the product against the success criteria defined during the inception phase.

These phases are mapped to a list of disciplines that include:

- Business modeling;
- Requirements;
- Analysis and design;
- Implementation;
- Test;
- Deployment;
- Configuration and change management;
- Project management;
- Environment.

There are a few other frameworks including DoDAF, GERAM, and the Nolan Norton Framework, but they have many similarities to the ones discussed so far. An architect should be familiar with these frameworks so that their taxonomy, methodology, and processes can be tailored to fit their enterprise. Next, we will focus on the modeling languages most commonly associated with the frameworks just discussed.

7.2   Architecture Modeling Languages

Many kinds of modeling languages can be used to describe the views selected when using a specific architecture framework. These languages may span one or more domain architectures and may be used to model the business and IT as well as providing traceability between domains. The modeling languages that we focus on are commonly used for different domain architectures and different industries:

- *IDEF:* IDEF is a set of modeling languages used mostly in systems and software engineering, and it covers a wide range of uses:
  - Functional modeling (IDEF0): Its core concept is the function, and this method refers to how to model the actors performing the function, the data consumed and produced, and the relationships between business functions.
  - Process modeling (IDEF3): This method aims to capture the flow of the business process via process flow diagrams, showing task sequence, and various different scenarios.
  - Data modeling (IDEF1x): This method focuses on creating logical and physical data models.

  The IDEF models provide ways to support different architectural viewpoints, but there is no relationship between them, so it can be difficult to create views that encompass multiple models and multiple architecture domains at the same time.

- *Business Process Model and Notation (BPMN):* BPMN provides a solid base for modeling business processes. It provides a notation for modeling these processes in terms of activities and their relationships, which is understandable by all business users. This helps to bridge for the gap between business process design and process implementation. BPMN is limited strictly to business process modeling and it does not cover the whole business architecture domain. The main types of models offered are process, collaboration, choreography, and conversation models. It overlaps with UML activity, flow, and sequence diagrams: they share common elements such as events (including start and end), activities (which in BPMN can be atomic; tasks and compound tasks can have subprocesses), gateways (equivalent with the UML decision object), swim lanes to organize processes, and artifacts to provide additional information.

- *Entity relationship diagrams (ERD):* ERD focuses exclusively on data modeling at three levels: conceptual, logical, and physical. The conceptual level contains the least granular detail, but establishes the overall scope of what is to be included within the model set. It normally defines master reference data entities commonly used in the organization. The logical level contains more detail than the conceptual model, because all the entities are defined, including the relationships between them, and because they are technology independent. The physical level provides the lowest level of detail, developed on the logical level, and contains enough data to create the database and be technology dependent. ERD's main principles include the entity (the thing that is recognized as being capable of an independent existence and which can be uniquely identified), attributes (properties of entities), and relationships (association between different entities). These relationships can have defined roles, as well as cardinalities. There are quite a few tools that can help a data architect build a data model at the required level of detail:

- Erwin: The most popular and comprehensive enterprise solution with a particularly powerful model manager for large teams.

- Entity Relationship Modeler: This helps you design complex database models for all major RDBMS vendors. Versions with the Forward Engineer feature can model entities and convert them into SQL Scripts, or even reverse engineer existing databases to visualize a database model. The convert feature can quickly convert models from one database vendor to another.

- Borland Together: This tool is mostly integrated with UML, but it also has ERD capabilities to forward and reverse-engineer.

- Toad Data Modeler: This modeler helps you create logical and physical data models and generate new database structures or make changes to existing models automatically.

- Enterprise Architect: It supports reverse engineering of a wide range of database repository schemas.

- Many other tools that are specific to certain database types exist, such as Oracle Designer, Power Designer (Sybase), and MySQL Workbench.

- *Unified Modeling Language (UML):* UML is a standard language used to produce application and technology blueprints. It can be used to visualize, specify, construct, and document the artifacts of a software-intensive system. It can be used to produce artifacts to document requirements, architecture, design, source code, tests, prototypes, and more. The core building blocks of UML include:

- The abstractions of the model can be structural. This mostly includes static parts of the model such as class, interface, collaboration, use cases, components, and nodes. This is also true for behavioral, dynamic parts such as interaction and states, as well as grouping parts (boxes into which a model can be decomposed) and notational parts (explanatory parts of the models).

- Relationships between parts can exist in several forms: dependencies (a change to one thing may affect the other), association (a set of links or connections between objects), generalization (objects of a specialized kind are substitutable for objects of a generalized one), and realizations (one object specifies a contract that the other one guarantees to carry out).

- Diagrams: A diagram is a graphical presentation of a set of things and their relationships that can help visualize a system from different perspectives. UML includes the following types of diagrams:

  - Class diagrams show a set of classes, interfaces, collaborations, and their relationships. They are the most common diagram found in object-oriented systems and they address the static design view of the system.

  - Object diagrams take a static snapshot of a set of objects and their relationships.

  - Use case diagrams can show a set of use cases and actors and their relationships. They address the static use case view of the system and are important in modeling the behaviors of the system.

  - Interaction diagrams display an interaction between a set of objects and their relationships, including the messages that may be dispatched among them. The two types of interaction diagrams are sequence diagrams (emphasizing the time ordering of messages) and collaboration diagrams (emphasizing the structural organization of the objects that send and receive the messages).

  - Statechart diagrams can illustrate states, transitions, events, and activities within a certain system. They emphasize event-ordered behavior. They address a dynamic view of the system and are especially important in modeling the behavior of interfaces, classes, or collaborations.

  - Activity diagrams are a special kind of statechart diagram that shows the flow from activity to activity within a system and are important in modeling the function of that system. Activity diagrams emphasize the flow of control among objects. They are also used together with use case diagrams to show the behavior of the system in terms of "happy path" (the default scenario featuring no exceptional or error conditions containing the sequence of activities if everything goes well), as well as failures.

- Component diagrams show the organization and dependencies among a set of components, addressing a static implementation view of the system. They are related to class diagrams, as a component typically maps to one or more classes.

- Deployment diagrams show the configuration of the run-time processing nodes and the components that live on them. They address a static deployment view of the technology architecture. They are also related to component diagrams, because a node contains one or more components.

- Common mechanisms are applied persistently across UML. Mechanisms include specifications (textual statement of the syntax and semantics of the building block), adornments (details like the type: abstract class, for example, visibility of the attributes and operations, and more), common divisions (including the class, object, interface, and implementation), and finally, extensibility mechanisms (making UML an open-ended language and making extension possible through the stereotypes, tagged values, and constraints).

- UML can define several views of the system architecture including:
  - Use Case view: Describes the behavior of the system as seem by its users and testers.
  - Design view: Encompasses the classes, interfaces, and collaborations that form the vocabulary of the problem and its solution and supports the functional requirements of the system.
  - Process view: Displays the application processes and threads that form the system's concurrency and synchronization mechanisms. It addresses the performance and scalability of nonfunctional requirements.
  - Implementation view: Includes all the components and files used to assemble and release a physical system. It addresses the configuration management of the system.
  - Deployment view: Encompasses the nodes that form the systems' hardware. It addresses the distribution, delivery, and installation of the parts that make up the physical system).

- *Systems Modeling Language (SysML):* The general purpose of this modeling language is to support the specification, analysis, design, verification, and validation of a broad range of systems. SysML is defined as an extension of the UML, and it uses UML's profile mechanism. However,

SysML adds two more diagram types (requirement and parametric) to assist with gathering requirements as well as performance metrics. It supports different kinds of allocations, including functional and structural. The models are designed to be exchanged using XML metadata.

- *Archimate®:* Archimate is a modeling language used primarily for enterprise architecture modeling. It focuses on the relationship between different layers of an architecture, and it offers insight into the alignment of business processes and their supporting applications, or the applications and their technical infrastructure. It is built primarily on UML by adding in the domain that was missing: business architecture. Therefore, the main layers identified in Archimate are:

  - The business layer, which offers products and services to external customers and that are realized in the organization by business processes (performed by business actors or roles);
  - The application layer, which supports the business layer with application services and that is realized by (software) application components;
  - The technology layer, which offers the infrastructural services (e.g., processing, storage, and communication services) needed to run applications and that is realized by computer and communication devices and system software.

  All of the layers specified above have different concepts and relationships structured in models, viewpoints, and views according to the ISO/IEC/IEEE 42010:2011 standard. Archimate was initially divided into three architectural layers (business, information, and technology) and three architectural aspects (active structure, behavior, and passive structure). With the introduction of a second version in 2012, the addition of implementation and migration added a fourth layer, while the motivation extension added an additional motivational aspect (for examples, see Figures 7.4 and 7.5).

  The most important concepts in Archimate, which can be found in more detail in the Archimate specification, are:

  - *Business:*
    - Business actor: Organizational entity capable of performing behavior.
    - Business interface: A point of service where a business service is made available to the environment.
    - Business service: A service that fulfills a business need for a customer.

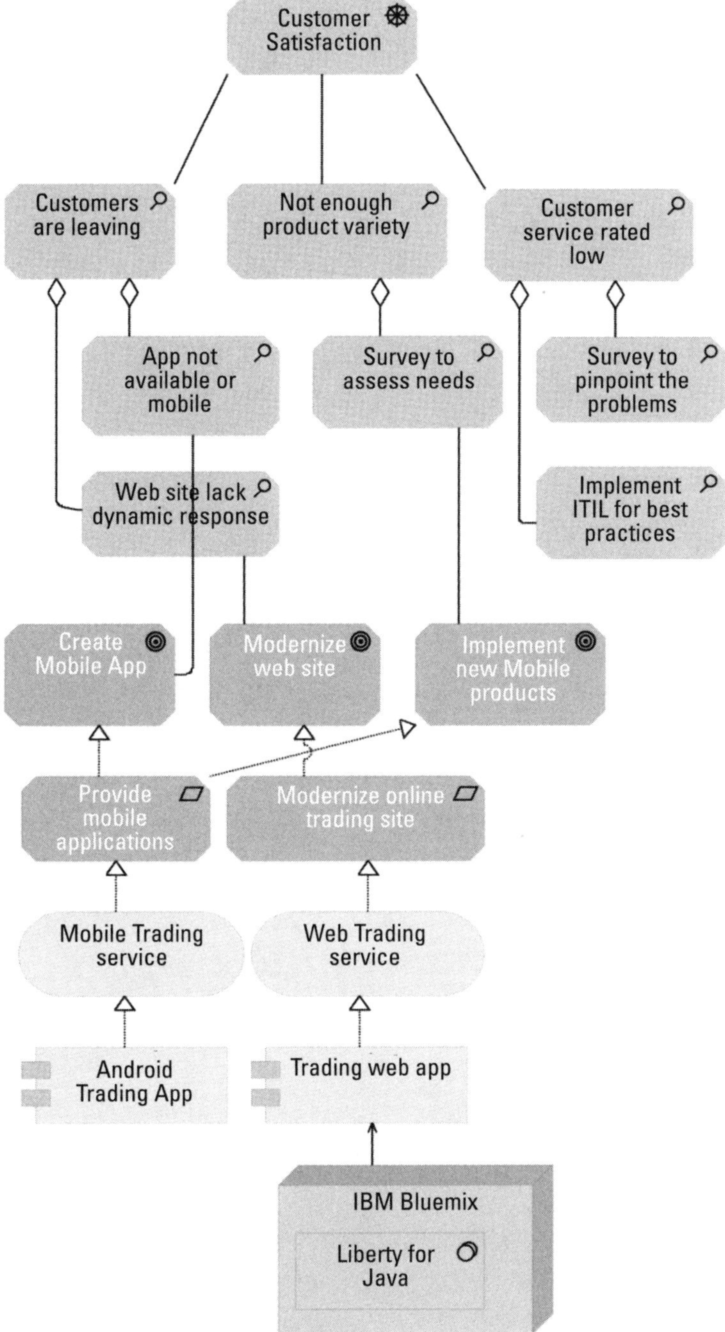

Figure 7.4 Example of Archimate traceability (using motivation package): customer satisfaction.

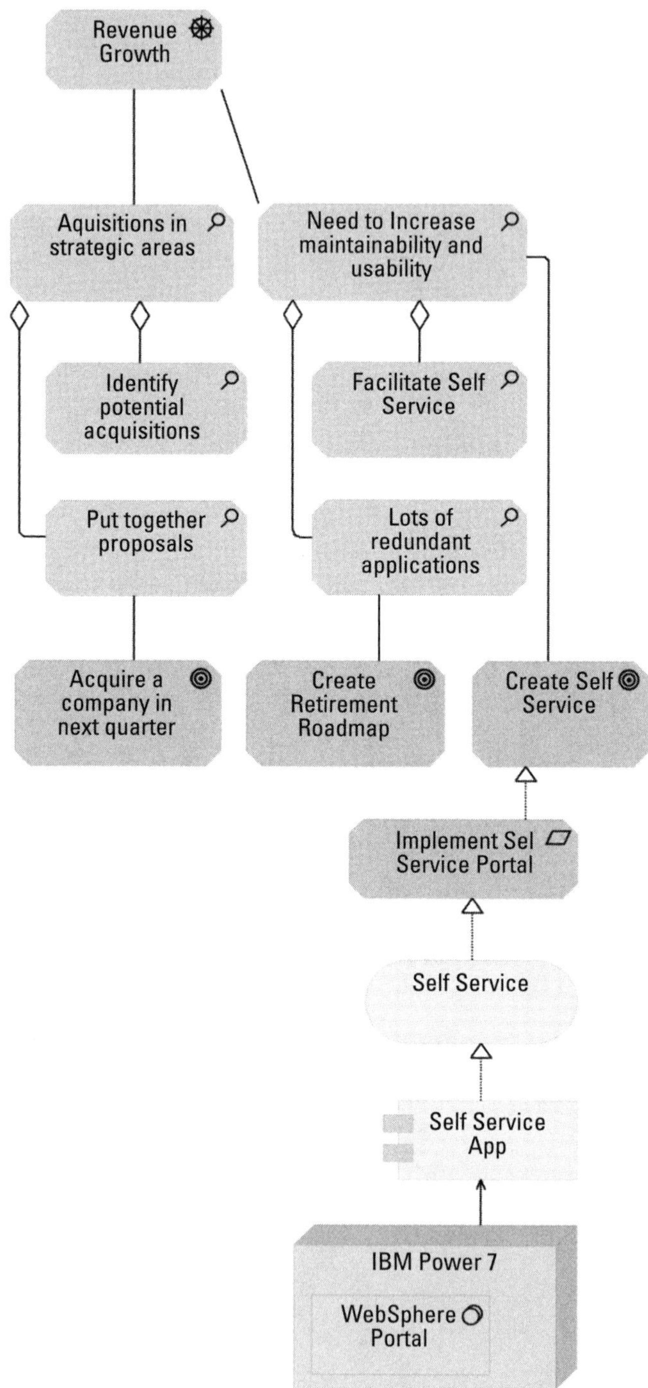

Figure 7.5 Example of Archimate traceability (using motivation package): revenue growth.

- Business process: A behavioral element that groups behaviors based on the ordering of the activities that produces a defined set of products.
- Business function: A behavioral element that groups behaviors based on the required business resources.
- Representation: Perceptible form of information carried by the business object.
- *Application:*
  - Application component: A modular deployable and replaceable part of a software system that encapsulates behavior and data and exposes them through a set of interfaces.
  - Application interface: Point of access where an application component is made available to a service or other application component.
  - Application service: Service that exposes automated behavior.
  - Application function: Element that groups automated behaviors that can be performed by a component.
- *Technology:*
  - Node: A computational resource where artifacts may be stored and deployed for execution.
  - Device: A hardware resource where artifacts may be stored and deployed for execution.
  - Network: Communication medium between two or more devices.
  - Infrastructure service: External visible unit of functionality provided by one or more nodes and exposed to well-defined interfaces.
  - Artifact: A physical piece of data that is used or produced in software development for the deployment of a component.
  - Infrastructure interface: Point of access where infrastructure services can be accessed
  - System software: A software environment for specific types of components that can be deployed in the form of artifacts.

Archimate's motivation extension also introduced concepts such as: stakeholder, driver, assessment, goal, requirement, constraint, and principle, used to model motivations behind the design or change of some enterprise architecture. This extension is a very good mechanism to model requirements and then to ensure traceability from the stakeholders' concerns and back to business processes and functions, and on to application services, components, and even down to the technology layer (infrastructure services, nodes, system software, artifacts, and more).

Once Archimate was adopted by the Open Group as TOGAF's architectural methodology, it became clear that, although it covered phases B–D and allowed modeling in many domains, the language still failed to cover any implementation and migration concepts. As a result, the implementation and migration layer was introduced in a second version of Archimate, which takes the work package as its central behavioral concept and which can produce certain deliverables (reports, software, and more). To describe the different stages in time for various domain architectures, the implementation layer introduced the concept of a plateau. It also serves to express the outcome of gap analysis between two different architectural plateaus.

Archimate also introduces a few standard viewpoints that can be used to describe how the stakeholder's concerns will be addressed by the architecture solution. Depending on the stakeholder, you might need to use one or more architectural layers and/or aspects. The business function and business process viewpoints show the two main perspectives on business behavior. The organization viewpoint depicts the structure of the enterprise in terms of its departments and roles. The information structure viewpoint describes the information and data used. The application, behavior, and cooperation viewpoints contain the applications, components, and their relationships. Finally, the infrastructure viewpoint shows the infrastructure and platforms underlying the enterprise's information systems in terms of networks, devices, and system software. The most complex viewpoint is the layered one, which shows traceability between the business actors/roles, services, and processes back to the application services and technology devices.

This can be quite daunting. Archimate uses a lot of concepts, relationships, extensions, layers, and viewpoints. The architect will need to know all of them so as to completely represent the enterprise architecture. What about each of the domain architectures? As mentioned in the previous section on TOGAF, the beauty of Archimate is that the architect can slice and dice. If you want to only describe the business architecture and to connect the business concepts with the stakeholder's goals and concerns, then only the business and motivation concepts are needed, and nothing else. If you are a systems architect and you only want to represent the application and technology layers, then you only need to choose your viewpoints that span these two architectural layers, and nothing else. As an enterprise architect, you should be able to coach the other architects to make sure they understand the enterprise architecture you put together and thereby help them to better create lower-level architectures.

7.3   How Do I Use All of These?

It is true that an architect needs to acquire copious knowledge to accomplish his or her work. Now we focus on some tips and tricks to help you master such architecture-related knowledge (frameworks, modeling languages, and the rest of the panoply).

The first thing to understand as a junior architect is that frameworks, methodologies, and modeling languages are only tools to help you do your job. An architect needs to be able to express his or her ideas using language that will be easily understood by all stakeholders, which means that:

- First and foremost, know all your stakeholders—and not just the decision-makers, such as account managers and sponsors. Consider everybody who has a stake in the project. If you are an application architect, for example, you also need to get feedback from, and sell your solution to, the lead developer, not just to the account manager. A systems architect would also need to take into consideration the vendors, infrastructure specialists, production managers, and other interested parties. Smooth implementations reflect well on you, and the best way to achieve this to make sure that everybody involved is on the same page.

- Once you know the stakeholders, understand their concerns. What kind of views should you produce to show that your solution will satisfy those concerns? This is the point where knowledge of frameworks, methodologies, and modeling languages becomes essential, because without them, you have no structured way to approach complex architectural problems at any level or domain.

In a way, it is like playing an instrument: you need to be able to play for different audiences, also known as stakeholders. How can one prepare to play the role of architect? Here are some tips:

- Make sure to select the right patterns, framework, methodology, and language for your kind of architecture. If you only work in infrastructure architecture, for example, then there would be little point in learning BPMN, or other tools related to business architecture. Start small by learning just what is of interest to your architecture domain. Over time, you will naturally add more to your knowledge base. If you are an application architect, UML is essential to start with, because you will need to model application services and components. Once you become a more experienced application architect however, you will start to learn more about the enterprise architecture frameworks and languages. You may still work on application architecture, but now you can use deliverables from the Zachman framework or the steps from Phase C (information architecture) in the TOGAF framework to come up with a better-defined application architecture. You will be able to understand the interfaces with other domain architectures. You may also learn more modeling languages or more application architecture frameworks, expanding your application architecture breadth.

- Adapt to the specifics of your job. Although there is some standardization among the many types of architects, it is nowhere near as well established as in other professions. The role of the architect can differ (sometimes quite a bit) from one company to another. A company might have its own methodology, specific deliverables, or processes: it will want you to use them. How do they measure the value of the architect? Is this just based on deliverables you produce and governance, or does this company value collaboration with other project members? As an enterprise architect, a company may only be interested to see business value in your design. As a business architect, you might have to interview business managers first, and use either a bottom-up or a top-down approach. As an infrastructure architect, this might mean that you will need to know specifics such as storage, networks, and so forth, in more detail. As an application architect, sometimes you may have to fill in the gap and perform code reviews, or act as a lead developer, making sure the development team gets all the information they need to deliver the needed functionality.

- Know the culture. Understand the specifics of the company represented by its internal culture. Both architectural knowledge and more informal information on the company culture, such as how mature the organization is, is there an internal methodology already in place, and so on, can be equally useful.

  One obvious example is knowing the attitude toward risk in your company, which can influence how a methodology or framework would capture such risks, gaps, and related mitigation/remediation actions. Workplace culture can definitely influence the patterns, governance, and the way you tailor frameworks to fit your organization. It may also influence the level of detail required for certain activities and deliverables from the framework and methodology that is adopted and tailored by the enterprise.

- Use the most appropriate tool for the task. The market is currently inundated with architecture frameworks, patterns, methodologies, modeling languages, and other tools. There is always a marginal advantage that one has over another. As the architect, you need to decide what is most appropriate. This also is true for each of the activities that you perform. For example, an application architect might use UML as the modeling language, but for the logical/physical data model, an ERD language and tool like ERWIN may be needed. If you focus exclusively on business architecture you can use BPMN, but if you want to show traceability, Archimate would serve much better. The reality is that there is no one-size-fits-all modeling language that can be used in any situation. That being said, some tools alleviate this issue by allowing you to use a large variety of modeling languages within the models you use them to build.

In summary, you should first understand the architecture needs that you must satisfy, and then select the most appropriate framework or modeling language. Approaching the process with a preconception that certain frameworks can only be used for certain companies, or certain languages can only be used for certain domains is not helpful. As with any other job, aim to keep yourself abreast of the latest trends so you are always able to come up with the right solution.

References

[1]   Gamma, E., et al., *Design Patterns: Elements of Reusable Object-Oriented Software*, Reading, MA: Addison-Wesley, 1994.

[2]   Clinger-Cohen Act (1996), Information Technology Management Reform Act, August 2006. http://dcmo.defense.gov/Portals/47/Documents/Clinger_Cohen_Act.pdf.

[3]   Sarbanes-Oxley Act (2002), http://www.soxlaw.com/, accessed May 2016.

[4]   TOGAF v9.1, Introduction, http://pubs.opengroup.org/architecture/togaf9-doc/arch/, accessed May 2016.

[5]   CMMI Maturity levels, http://www.tutorialspoint.com/cmmi/cmmi-maturity-levels.htm, accessed May 2016.

[6]   Magic Quadrant for Enterprise Architecture Consultancies, http://www.ey.com/Publication/vwLUAssets/Magic_Quadrant_for_Enterprise_Architecture_-_24_June_2015/$FILE/Gartner%20magic_quadrant_for_EA%20Consultancies.pdf, accessed May 2016.

8

Deliverables Produced by an Architect

Architecture deliverables or documentation establish documenting and communicating the model used to understand the enterprise, system, application, and network through a series of views (based on predefined viewpoints) as well as the architectural thinking behind the solution architecture. Depending on the type of architect, different documents might be required, and deliverables might also be recommended in keeping with the architecture framework or methodology adopted by an organization. More information regarding the contents for each of these specific deliverables can be found in the documentation for the individual frameworks and methodologies; this chapter focuses on the types of documents that map to the architect's fundamental responsibilities. The most common input into such architecture documents will be reviewed, as well as architecture thinking and how architects extract information and use it as an input into that architectural thinking.

8.1 Input for the Architecture Deliverables

8.1.1 Project Charter/Request for Architecture Work

As an architect working in a project to implement a specific solution, or in an architecture initiative trying to define the work packages based on the business goals and drivers, initiative is needed to provide a preliminary delineation of roles and responsibilities, objectives, and identification of major stakeholders. This is usually called a project charter, project definition document, or request for architecture work and serves as a reference for future changes in the project/initiative (see Section 8.1.4). It includes the reason for implementing the project, its sponsors and major stakeholders, directions concerning the solution, in-scope and out-of-scope items, time limits, high-level communications, management plan and high-level budget, financial constraints, organizational and business constraints, and changes in the business environment, as well as external and regulatory constraints.

Example 8.1: Table of Contents Example for a Statement of Work Document

Table of Contents
1. Executive Summary
2. Version History
3. Background
4. Purpose
5. Project Request and High-Level Requirements
6. Project Description
7. Overview of Vision
8. Approach
9. Change of Scope Procedures
10. Roles, Responsibilities, and Deliverables
11. Acceptance Criteria an Procedures
12. Project Plan/High-Level Schedule

Example 8.1 shows a table of contents from a roadmap statement of work (SOW). The overview of the vision (item 7) is usually a conceptual diagram that depicts the core concepts, and all participants are mapped into a RACI matrix. RACI is an abbreviation for four key responsibilities associated with projects, and identifies who is Responsible, Accountable, Consulted, and Informed about specific tasks or actions.

8.1.2   Requirements

To know how the enterprise, system, and application are required to behave to design a solution, the following items or elements must be considered:

- List of the stakeholders (including the sponsors of the initiative) and their interests. Based on the importance of the stakeholders, their requirements may be prioritized and some of them deferred for future initiatives. Most of the time, these stakeholders are business or IT leaders, account managers, and operation managers. Each of these might have different requirements on how the business, system, or enterprise should work to provide the maximum benefit from the business, IT, or operations point of view.

- In-scope and out-of-scope requirements to define the boundaries for the initiative.

- Constraints (based on the type of the initiative) including organizational and budget/financial, external, business and enterprise architecture, and time.

- Assumptions made that are ideally fully validated once the document is in a final form.

- Needs and features as an initial, unstructured set of requirements gathered during interviews with the stakeholders.

- Analysis process: how needs and features are transformed into actual requirements.

The main types of requirements that should be captured to ensure a solid base for the architecture design, as well as the other SDLC phases are:

- *Functional requirements:* Define the functions of the system or application. The input for these is the needs and features document. Features represent initial unstructured requirements as gathered from the participating stakeholders (initial interviews). Needs are part of the problem space, while features are part of the solution space. There are a few methods that can be employed during the analysis process, including use case diagrams with successful and unsuccessful paths (mostly used within the projects) and business scenarios used to document the business problem, such as the business and technical environment, the objectives, and the actors with their roles and responsibilities. The most important of these is the business value of solving the problem (used for business and enterprise architecture initiatives), as well as different templates to ensure that the requirements are as complete as possible (such as the Volere requirements specification template). These requirements usually contain the following components:

  - Problem statement as the jumping off point for requirements that should include a description of the problem, the impact of not solving the problem, stakeholders affected by the problem, and what a successful solution look like.

  - Stakeholders' profiles: describes each potential stakeholder for the system by gathering information about their type, responsibilities, involvement, and importance.

  - High-level business model: provides a business model context for the requirements including the major business processes and functions.

- *Requirements analysis:* this might include use case diagrams, happy path, and other useful and illustrative documents (please refer to the UML section above).

The following are separate subtypes of functional requirements that try to address different views:

- Business requirement: A goal of the organization requesting the system.
- User requirement: A task that the user must be able to accomplish using the system.
- Data requirement: Information the system or user requests/provides to satisfy an interface requirement or functional requirement.
- System requirement: A minimum set of hardware or other software components that a specific application component needs to run optimally.
- Regulatory requirements (also known as constraints): The requirements that a certain system or application must support to comply with all applicable laws and regulations.
- Design requirements: The requirements that the system or application must conform to from a design point of view. These are normally captured as constraints in the architecture document, although some of them might also be included in the functional requirements. Some examples might be using a certain application architecture framework, or an application server such as WebSphere.

All these types of requirements are normally included in the functional requirements document and should all be addressed in the architecture document.

- Nonfunctional requirements: A special type of requirements specifying criteria that can be used to judge the operation of a system, rather than its specific behaviors. This document is one main input for the architecture document. They are also known as engineering or quality requirements because they refer either to execution qualities such as security, performance, and usability or to evolution qualities such as testability, maintainability, extensibility, scalability, and extensibility.

These requirements specify how the system should perform and operate, with a primary focus on data (purging, backup), security (authentication, authorization, and more), performance (how the system should perform), and availability and maintainability of the system. The architecture document will take this as an input and define how the solution satisfies these requirements.

Proposed templates for the functional or nonfunctional requirements are available on the Internet. A few useful Web sites that have such documentation include:

- Smart BA gives examples of how requirements should be documented, as well as negative examples of poorly documented requirements and common mistakes made with requirements [1].
- Ofni Systems offers specific examples of functional requirements, the requirements traceability matrix, and other validation document resources [2].
- Functional and nonfunctional requirements document example [3].

8.1.3 Business Principles, Goals, and Drivers

Primarily for the business and enterprise architecture, business principles, goals and drivers define the context for architecture effort. These can significantly impact the way that an architecture is developed and will shape the business architecture as well as other domain architectures including application, data, and technology.

Some of the most widely used business principles include: clients come first, business integrity, fair competition, and permanent compliance with regulations. More examples in this vein appear in Eric Jacobson's 2010 blog post "Good Sample Business Principles" [4] and on the Web site of the International Chamber of Commerce [5].

8.1.4 Change Request

As more facts become known, or external disruptions (such as market factors, changes in business strategy, and new technology opportunities) open up opportunities to extend and refine the architecture, it is possible that one's current project charter might not be suitable to complete the implementation of a solution. As a result, the project can either deviate from the suggested architectural approach or request a change in scope and timeline.

A change request may be submitted to request a change in the timeline as well as budget for the specific project or architecture initiative. This would include a description and a rationale for the change, as well as references to the specific requirements that led to the change request. An architect would normally document the high-level technical requirements and a solution, recommending proof of concept initiatives and approaches.

8.1.5   Tailored Framework and/or Methodology

Chapter 7 discussed various frameworks and methodologies and how each is used for specific types of architecture. Each framework and methodology can be tailored to meet the specific needs of a company or organization. These needs might include integration with other frameworks and methodologies such as project management frameworks, internally grown frameworks, and change management frameworks. The formality and level of detail can also be tailored to align themselves with the culture and existing levels of architecture maturity.

The content of various architectural deliverables might also be subject to tailoring. In many companies, one comprehensive template covers all possible cases and sections that apply to the task at hand just need to be selected. For example, this might entail a solution architecture document template. If there are no changes at the data level, there need not be any data view such as the data entity/application function matrix or the data dissemination diagram. Similarly, a system architecture document that has no infrastructure changes can omit the deployment view that is used to depict the way applications get deployed in the enterprise.

8.1.6   Capability Assessment

It is important to understand the various levels of capabilities involved, such as capability of the enterprise as a whole, the maturity of the IT function within the enterprise, and the maturity of the architecture function within the enterprise. Various aspects related to the activity include answering these questions:

- What architectural assets are currently in existence?
- What standards and reference models need to be considered?
- Are there likely to be opportunities to create reusable assets during the project?
- Is the readiness of the business to transform itself to reach the target level of capability: the current target readings and associated risks?

8.1.7   Organizational Model

This shows how the enterprise is organized as well as the main roles and responsibilities of the groups and individuals involved. It is of particular importance for the business and enterprise architects to understand this model so they may correctly identify all major stakeholders, prioritize their requirements, layer financial constraints and investments, and show which groups are affected the most.

8.2 Architecture Deliverables

The types of deliverables that an architect would produce depend to a large extent on the type of architect who is doing the work. There are always different templates and special architectural deliverables specific to a company. However, as with architecture frameworks, an architect is required to customize both the template and its content to meet the needs of your stakeholders.

8.2.1 Architecture Document

One of the most important documents that each architect must produce is the architecture document. Depending on the type of architect, this might be called a business, application, infrastructure, system, or solution architecture. Its purpose is to describe a high level of design so that it can be understood and validated by stakeholders and guide more detailed design activities that will eventually lead to the implementation of functionality that it proposes.

The following information should be included in such a document:

- *Executive summary:* This is usually the first section in order, but the last one to be completed. The summary contains a brief overview of the most significant aspects of the architecture and should be able to stand on its own. It should serve as a high-level document for management or executives who need to understand the design and its impact as it relates to other business processes, systems, applications, or infrastructure components.

Example 8.2: Executive Summary

This document contains the architectural approaches and decisions for the ABC Web application. This application is being designed and built to serve the current and future needs of the enterprise. The key objectives are content by the business and a more robust, scalable, and extensible technical foundation. This document provides a high-level description of the solution architecture required to meet the functional and nonfunctional requirements. Included in this document are specifics related to the system context, technology stack, deployment approach, application framework, technology architecture, and static and dynamic data approach and considerations.

This document is organized in a logical flow to describe the solution motivation, current state, and target architecture as well as architectural risks and solution rationale.

It is important to note that the purpose of architecture is to provide a description and approach for the core foundation of the Web application and will not focus on the detailed design, which includes patterns and implementation details of the application framework.

- *Supporting documentation:* This is the section that should list all input deliverables, including the requirements documents and their location(s).

- *Background:* This section provides information about the history of the current state of the application, system, and so forth.

Example 8.3: Background Section

Issues with the current state that need to be addressed to meet the business needs include:

- There is a disjointed and inefficient user experience due to the way that the various sections of the site are managed and owned.

- Technologies in the current state are old, not according to the technology standards (most of the case unsupported).

- The current Web site cannot be used on the mobile devices; it needs to be designed with dynamic responsiveness as a base principle.

- The analytics, mobile strategies, and user experience were not considered during the initial design of the Web site.

Therefore, the role and skill gaps with respect to addressing and updating the necessary components of the solution need to be considered as part of the people dimension of the design.

- *Constraints:* This section provides additional or summary information about official constraints within which this architecture must work or be designed.

- *Assumptions:* This section provides additional or summary assumptions upon which the draft architecture is based. Ideally, all assumptions should be validated before the final version of the architecture document is published.

- *Enterprise architecture directions/guidance:* This section includes the key guidance principles that shape the solution architecture including:

  - Build solutions based on latest industry best practices to ensure getting things right the first time.
  - Build white-labeled solutions for extensibility.
  - Maintain alignment with other strategic enterprise initiatives.
  - The application should be built on HTML5 (following responsive principles as required) to support Web and mobile devices.

- *Current/target architecture state:* This is the most important section of the document because it contains the views (based on various viewpoints) that show how the architecture models created for the business, data, application, or technology domain architectures will satisfy stakeholders' concerns. The important thing is to select the most relevant viewpoints on the basis of the stakeholders and their concerns, create the current (or baseline) state and then the target state, do the gap analysis to come up with the roadmap components, make sure there are no impacts to other projects going on in parallel, and then review the architecture with the stakeholders. Depending on the type of project or initiative, there may be multiple transition (or project) architectures that will show how the business, application, infrastructure, or system will evolve from the current state to the target one, by going through a set of intermediate steps. At this point, the business, application, and infrastructure capabilities will be broken down into realizable increments that can be included in separate projects (refer to Figure 5.5), which showed how transition architectures will get the system from its current state to the target state). The section will include separate views created from the same model, but according to various viewpoints mentioned earlier. Various architecture frameworks or modeling languages have different viewpoints along with some guidelines of when each of them should be used. These viewpoints might be restricted to one domain architecture or they might span multiple domains. TOGAF, for example, has a multitude of viewpoints classified in catalogs, matrices and diagrams as shown in Figure 8.1.

Another example is Archimate. It is a modeling language, but it also has standardized viewpoints; some of them are restricted to one domain architecture (like business function or process, application structure, and application behavior), but some span multiple domain architectures (like the *service realization* viewpoint used to show how one or more business services and processes are realized by underlying application services and components, or the *implementation and deployment* viewpoint that shows how one or more applications are realized by the infrastructure services and components). The most complex viewpoint is the layered viewpoint, which provides an overview of (a part of) the business architecture going down to the infrastructure architecture in a single picture, showing the impact of change for extending the business with new services. The example shown in Figure 8.2 comes from the banking industry and it goes through each of the domain architectures.

- At the business architecture level, roles and actors use business services (like *provide mortgage quote service*) that are realized by business

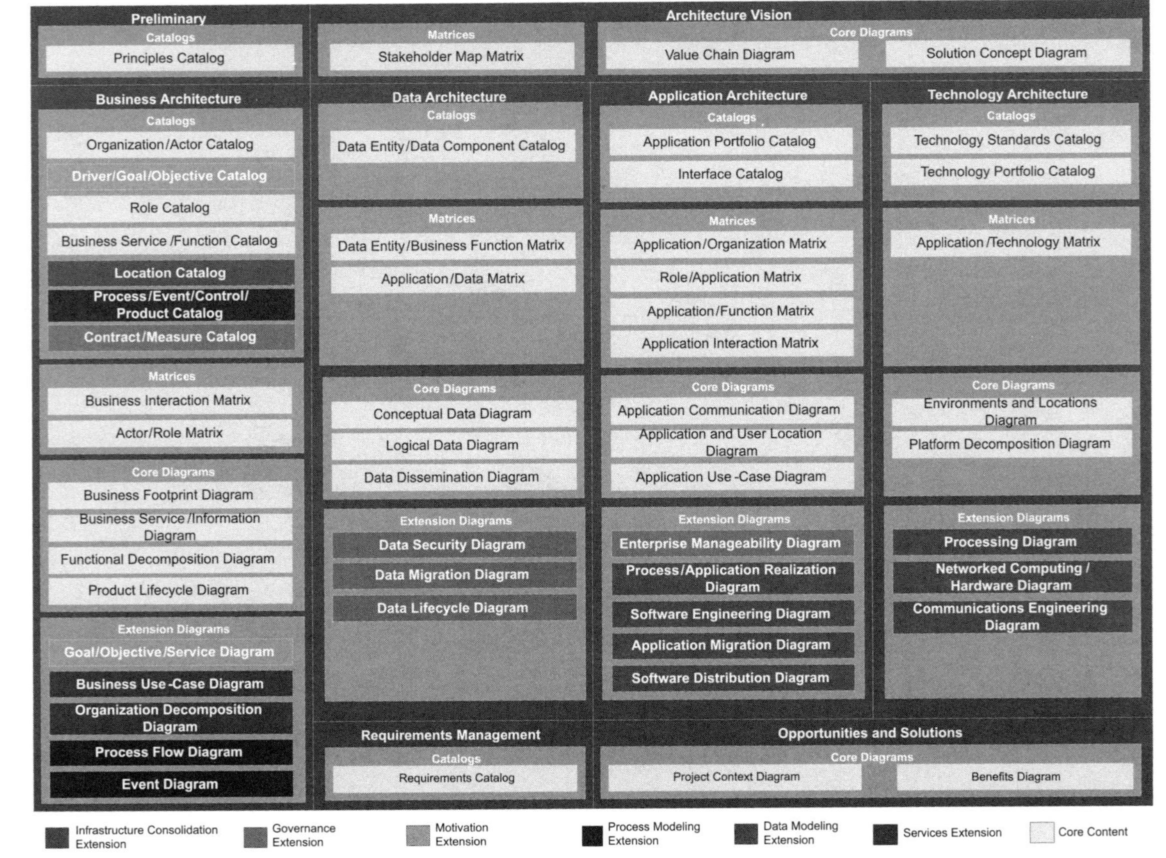

Figure 8.1　TOGAF Viewpoints. (Source: TOGAF® specification.)

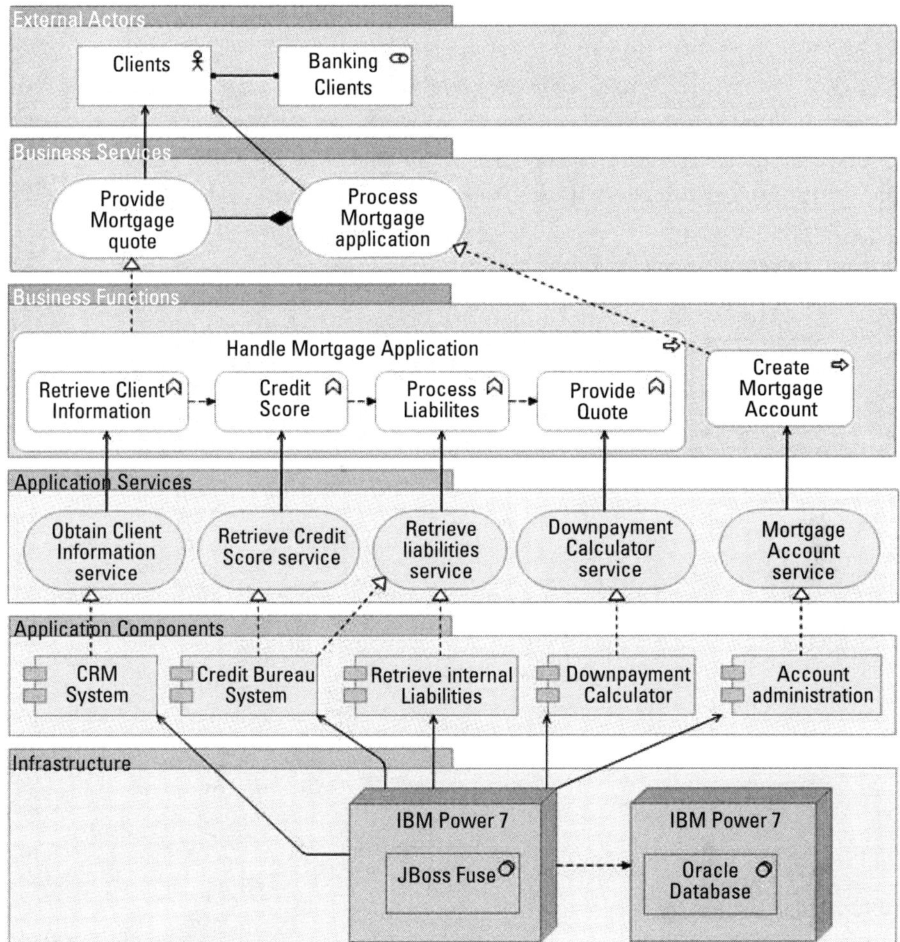

Figure 8.2   Archimate layered viewpoint.

processes and functions (*get client information, get client score,* and so forth).

- Going into the application architecture layer, we can see that the business processes use the application services (*obtain client information, retrieve credit score, retrieve liabilities service, downpayment calculator service,* and so forth), which are realized by the application components grouped into systems customer relationship management (CRM) system, *credit bureau system, account administration*).

- At the infrastructure architecture level, application components are implemented on nodes (IBM Power 7), which contain various system software packages (*Jboss Fuse, Oracle Database*).

The example is centered on one main business process and can be used to show what underlying services and components serve it.

No matter what modeling language or architecture framework you use, the point of this section is to clearly depict changes at the different architecture domains that will be introduced by the target state. This might include new application components or infrastructure servers, new services, and so on.

In Figure 8.3, the existing infrastructure is enhanced by adding a new WAS (WebSphere Application Server), as well as a new Oracle Server, and clustering these together to improve system performance. The ar-

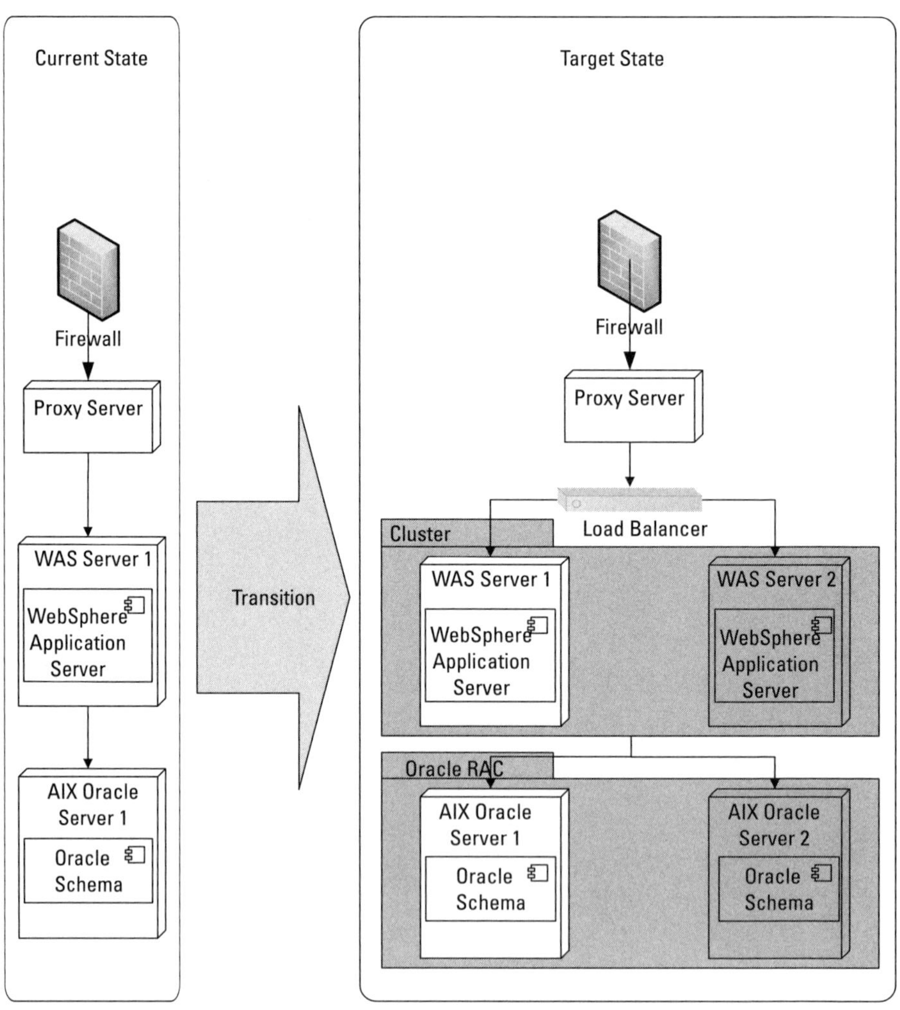

Figure 8.3   Example of current versus target state infrastructure view.

chitecture document should describe these new components and how they support certain requirements (see Table 8.1).

- *Architectural risks:* This section of the report focuses on the most significant risks, from the architecture point of view (not from the initiative, project, or program point of view). Certain risks could still remain if there are constraints that prevent the design and implementation of a system that meets the requirements with complete certainty. This section should describe the risk, the impact, and the probability of it happening, as well as measures to reduce the impact if this event occurred (please refer to Table 8.2). Also, consider ways to reduce the probability of this event occurring within the expected lifetime of this system, together with a contingency plan for how to react if the risk occurs. As an example, consider the residual risks that might exist because of the engineering properties of the architecture:

  - Time or funding constraints might have prevented a thorough proof of concept in terms of the initial system's ability to provide acceptable performance.
  - There might be some doubt regarding the system's ability to scale over time and support projected traffic or usage volumes.

- *Architecture decisions:* This section of the architecture document provides a rationale for the architecture design. It should document decisions made to establish its architectural characteristics. First, clearly describe each issue that required making a decision. Include all alternatives considered, along with the decisions eventually reached and the rationale behind them (please refer to Table 8.3). Each decision might require a separate document to describe the issue in detail, alternatives considered, and criteria used to make the decision.

Table 8.1

New Components

Component	Description	Requirement
WebSphere Cluster	WebSphere application cluster provides capabilities of high availability by providing fail over and load balancing mechanisms.	Web site performance: customer record retrieval should not exceed 2 seconds.
Oracle RAC	Oracle RAC uses Oracle Clusterware for the infrastructure to bind multiple servers so they operate as a single system.	
Load Balancer	Distributes workloads across multiple computing resources and aims to optimize resource use, maximize throughput, minimize response time, and avoid overload of any single resource.	

Table 8.2
Application Architectural Risks Example

ID	Description	Imp. (H/M/L)	Prob. (H/M/L)	Risk Reduction	Contingency Plan
1	Use HTML5 tags that are unsupported by the system	H	M	Installing patches to resolve the issue	Avoid using the offending HTML5 tab. SA and Lead Developer to work to find an alternative
2	Plug-ins required for the application may experience defects and/or compatibility issues	M	L	Engage consultants with experience for the specific application framework.	Find alternative plug-ins that meet the requirements or other custom alternatives.

Table 8.3
Application Architecture Decision Examples

ID	Issue	Alternatives	Decision	Rationale
1	Decide what deployment configuration to select for the application	(1) Deploy in the same Weblogic cluster as the DEF application (2) Deploy the application in a separate cluster	Deploy the application in the same cluster.	Follows the deployment design pattern for the other applications deployed.
2	Choose the tool for PDF generation	(1) Use Phantom JS (2)Use Jasper Reports	Use Phantom JS	Respects the requirement to have similar look and feel with the main site. Using the same tool for PDF generation as the one used for the Web sites ensures consistent look and feel.

- *Architecture gaps:* Ideally, this section should be empty, because it indicates any gaps left between requirements for the solution and the actual solution to be implemented.

- *Future improvements:* This section should contain any improvements that could be made to the current architecture not included in the current solution. It should also explain why desired future improvements were not included in this proposed architecture, as well as a brief justification as to why these would be useful.

Depending on the various templates different companies require, other sections might be included in the architecture document, which sometimes might detail the intersection between architecture and infrastructure, architecture and business design/modeling, and so on.

8.2.2　Architecture Decision Document

The architecture decision document helps to maintain a record of important decisions made during the development of the broader architecture. The architect produces this document to help to evaluate solution alternatives with the objective of setting architectural direction for a project, business, or enterprise. It contains the architect's recommendations for specific alternatives together with the reasons behind them. It can be used for multiple purposes, such as:

- Clearly defining the problem statement, the relevant stakeholders, and their concerns.

- Defining the business requirements and constraints that must be addressed or accommodated.

- Providing a way to communicate decisions to key stakeholders and gain consensus or approval for portions of the systems architecture, without having to document the entire architecture.

- Delivering architectural guidance to the project team and stakeholders by providing a basis for the architectural direction as the project life cycle progresses.

- Document the decision for future architecture reviews or changes to the system.

As with other architecture documents, an architecture decision document can be created at various levels such as project, system, and enterprise. The final recommendation might document certain risks and implications that need to considered. For example, new technologies introduced locally, or within the enterprise, might require a proof of concept to validate engineering dimensions.

For small architecture decisions, there might be no need for a separate document if an architecture document already exists. Smaller architecture decisions can be documented in the architecture document if the recommendation was clear, and there was no need to spend time weighing alternatives. For a project where there is an architecture document, as well as one or more architecture decision documents attached, those documents should be referenced in the architecture document so a complete picture of the architecture is represented.

An architecture decision document usually contains the following sections:

- *Executive summary:* Similar to the architecture document, it should begin by summarizing the contents for high-level management or executives. The most important subsections to include are the problem statement and the recommendation.

Example 8.4: Architecture Decision Document: Executive Summary Section

1 Executive Summary

A fully functional disaster recovery solution for both lines of business will require the gateway systems to be designed, implemented, and deployed to close the disaster recovery gaps identified:

- Bidirectional file transfer between system AAA and system BBB;
- Real-time online transfer;
- Break sheet reconciliations;
- Upstream and downstream network connectivity and file transfer.

The required server infrastructure, network components, and telecommunication circuits will be acquired and deployed at the disaster recovery site to establish the run-time environment.

1.1 Problem Statement

The ideal pattern for the disaster recovery architecture is to create a one-to-one relationship between the production and disaster recovery site architectures to minimize functionality changes and costs and maximize reliability by providing the simplest procedures to activate disaster recovery systems in the disaster recovery declaration scenario. The one-to-one replication pattern relies on a production identical or similar infrastructure components at the disaster recovery site. The gateway of the AAA system relies on an Oracle RAC that does not exist at the disaster recovery (warm replication model).

This problem is specific to the AAA system and not applicable to the BBB one. This document will only detail the line of business 1 portion of the disaster recovery architecture.

There are four options that were considered in this document and for two of them a proof of concept was conducted.

- *Introductory sections:* As with the architecture document, architecture decision documents will have a few introductory sections, including background (current state of the process, application, or system related to the solution that is under consideration), requirements (short summary of the most important functional and nonfunctional requirements that drove the need for the decision), assumptions (on which the recommendation is made), and constraints (to which this architecture decision is bound).

- *Outstanding questions:* This section contains any questions still outstanding left unanswered while working through alternative analyses. This includes questions not answered by requirements and not dealt with by assumptions or constraints.

Example 8.5: Outstanding Questions

Outstanding Questions

Initial conversations with business and technology stakeholders have provided cautious resistance to leveraging the infrastructure in the disaster recovery warm zone. The current architectural pattern is to view the disaster recovery zones for line of business 1 and line of business 2 as separate and isolated. As the core systems become more integrated in production, the question that has to be answered is if the principles/patterns for separation have to change.

Sharing the systems and networks at the disaster recovery center has to be carefully investigated (proof of concepts required). There are possible risks and complications with the integration as the overall assumption is that the need for disaster recovery for the various lines of business is considered independently. Some shared services may not be available, which might be considered critical by the business.

- *Options/alternatives:* This section of the architecture decision document lists the various options/alternatives and provides any pertinent information about the solution including nonfunctional characteristics, system interfaces, audit capabilities, security, disaster recovery, and any additional data relevant to understanding and comparing alternatives. Here the architect should document the pros and cons, as well as the implications of recommending a specific alternative. This section will contain not only viable alternatives actually considered as potential solutions, but also alternatives discarded earlier. For the latter, the architect should also specify the reason why those solutions were not ultimately considered.

Example 8.6: Template for Documenting Alternatives

Alternatives
Alternative 1: [Short Name]
        [Description]
Alternative 2: [Short Name]
        [Description]
And so on
Alternatives not pursued
        [Short Name of alternative 1 not pursued]
        [Reason}
        [Short Name of alternative 2 not pursued]
        [Reason}

- *Analysis:* This section presents a dashboard to compare viable alternatives based on a list of preselected criteria that might include (but are not limited to):
  - Nonfunctional criteria such as availability, maintainability, performance, recoverability, security (authentication, authorization, and more), and adaptability. These criteria can be a subset of those recommended in your preferred architecture framework.
  - Business impact, including key business drivers and compliance.
  - Implementation impact, including budget, scope, quality, and delivery risk.
  - Maintenance impact, including operational cost, technical risk, organizational compatibility, and direction for the solution.
  - Architecture compliance, including the compliance criteria for the various infrastructure, applications, and data (please refer to Table 8.4).
- *Recommendation:* This includes the final recommended solution for the problem. It should detail the rationale and highlight key advantages to the recommended alternative over others considered and rejected. This section should also include the various risks associated with the proposed recommendation, an estimation of its impact and probability, risk reduction measures, and a risk contingency plan.
- *Various appendices:* These can include any other relevant information such as glossary of terms, detailed plan and costs, decision log, links to

Table 8.4
Criteria to Be Used When Analyzing Alternatives (with Examples)

	Option 1: Microservices Deployed in the Docker Container	**Option 2: SOAP Web Service Deployed in JBoss Fuse**
Business strategies		
Business risks		
Compliance		
Budget		
Scope		
Quality		
Schedule	This is the first implementation of microservices in the organization; best practices and patterns need to be developed.	Schedule will not be impacted as the organization has already best practices and patterns for SOAP implementations in place.
Delivery risk		
Technical risk		
IT operational cost		
Business operational cost		
Technical risk		
Training	Requires training as the development team does not have the necessary knowledge	No extra training is required as the development team has extensive experience JBoss Fuse.
Solution direction	This option is in line with the mid-tier strategy closing the gap to achieve the target state.	This option maintains the status quo and does not close the gap to achieve the target state
Organizational compatibility		
Availability		
Maintainability	Easier to maintain as a defect or a functionality change affects a limited number of services.	More difficult to maintain as all the services are incorporated in one EAR file which has to be redeployed every time a change is implemented
Manageability		
Performance	Microservices are light services that can easily handle an increase in workload.	More work is required to properly adjust the performance of the SOAP services.
Reliability		
Recoverability		
Component		
Disaster		
Accessibility		
Adaptability		
Interoperability		

Table 8.4   (continued)

	Option 1: Microservices Deployed in the Docker Container	Option 2: SOAP Web Service Deployed in JBoss Fuse
Scalability	Microservices scale very well, especially when deployed in a virtualized/cloud environment. New instances can be started in seconds	Not so scalable and flexible, formal verbose contracts, state full with bigger payload, does not scale so well.
Portability		
Extensibility	The functionality can be easily extended as each piece of functionality is encapsulated in a separate service.	Extending the SOAP service and adding more methods requires redeployment of the whole ear file and regression testing.
Assurance		
Auditability		
Security	Requires more work to secure the services including the communication between them.	SOAP has protocol extensions WS* to ensure the security of the information transmitted.
Privacy		
Integrity		
Credibility		
Usability		

additional information in regards to the options, vendor reference materials, and project documentation.

As with any other deliverable, an architect should try to keep architecture decision documents lightweight and practical, while still providing all necessary context. Other people who might contribute to an architecture decision document are project or initiative stakeholders, as well as any other architects who might have a role in architecture governance and in ensuring that the proposed recommendation will fit into the big picture. For example, there might need to be a decision made about a certain technology or vendor product to be used in an existing application or used to automate a certain business process. In this case, the systems architect should advise because that domain is where the technology or vendor product will be used. Keep in mind that these are only the most common sections included in such a document, but that additional information might be needed depending on what type of architecture is involved. Relevant resources discussing the purpose and content of these documents can be found in [6, 7].

8.2.3   Implementation Strategy

The architect is involved in a project throughout its whole life cycle, from the early phases when the scope is first defined, all the way through to the final implementation stages. The architect should continue to provide architectural governance throughout its implementation. An implementation strategy document should be created to provide a schedule for implementation and guide that process through predefined transition architectures. It should focus purely on the strategy that should be employed to get from the current state to the target state. The implementation strategy document is usually written in the later phases of the project, because elements of the architecture document will need to be outlined beforehand, including the baseline and target state of the various domain architectures (business, data, application, technology). The implementation strategy document should identify the gaps between current and target states. The impact of the interoperability requirements on the implementation must be evaluated, and the risks of and readiness for implementation must also be assessed.

As other architecture documents, an implementation strategy document contains introductory sections, which include an executive summary, a background, an intended audience, and supporting documentation. Sections more specific to the way that architecture changes are implemented include:

- *Strategic implementation direction:* This direction is based on the capacity of the enterprise, or line of business, to absorb the change already architected. Depending on the type of architecture involved, the content of this section can vary. For enterprise architecture for example, you might need to choose between the Greenfield, Revolutionary (radical change), or Evolutionary (phased/incremented change) approaches. The most common is the evolutionary approach, where changes required to push an enterprise to the target state are introduced sequentially. For application architecture, the implementation strategy should also include how the implementation will happen (phases, iterations, and so on), and any required conversion or migration strategy. This section should also present a disaster recovery strategy and how this could be implemented at a disaster recovery site. The main concerns with regards to disaster recovery are:
  - State infrastructure/application recovery time and recovery point objectives (RTO/RPO). Assess the impact of this change to the current RTO/RPO.
  - Assess if the application recovery architecture needs to be revised, or if a new plan is required. This includes any new infrastructure needed at the disaster recovery site to support the change in the functionality.
- *Implementation approach:* This section addresses how the strategic direction will be implemented. The most common approaches are quick wins, achievable targets, or the value chain method. The selected ap-

proach and dependencies shape the creation of any transition architectures.

- *Identify and group work packages:* At this point, the architect should group architectural activities logically, using the consolidated gaps from the various domain architectures. Activities should be separated into work packages that will deliver a discrete outcome. Not all outcomes will produce a direct business outcome. If there is a business outcome associated with a work package, it must have a clear business value associated. Each of these work packages will span two transition architectures (refer to Figure 5.5). The following activities should be taken into consideration for each of these work packages:

  - Impact on existing systems: Document the impact of changes on all interfaces, from business or technical systems, as well as downstream systems.
  - Conversion strategy: This may be required where the scope covers any conversions, including but not limited to data conversion from one system (usually a legacy system) to another.
  - Technical training strategy: Each of these work packages can potentially introduce new technologies or products. The purpose of this section is to identify the nature of the training required (for example, need spring training for developers). Include the version of the product, what particular group of people should be involved in training, and a timeline (when this training is required).

- *Group work packages into projects and prioritize them:* Prioritize projects by assessing the business value they delivered, set against the cost of delivery. The approach is to first determine, as clearly as possible, the net benefit for all of the architectural building blocks these projects will deliver. Then, verify that the risks have been effectively mitigated and factored. Prioritization criteria should include key business drivers as well as cost reduction, consolidation of services, and imposing a minimum number of transitional states.

As with other forms of documentation, implementation strategy may be needed for different levels of architecture. At the enterprise architecture level, work packages will include changes for all domain architectures starting from business and going all the way through to application and technology. For the changes that are confined within one system, the implementation strategy will include work packages with a scope limited to the application or technology domains. As for application or infrastructure architecture, work packages might only include changes in these two domain architectures. In this case, additional sections might be valuable, such as data conversion, or decommissioning of a legacy application or piece of infrastructure. The document should be tailored to the specific needs of the architecture being developed, keeping in mind that each section that does not provide value, or that can be referenced from another document, should be either eliminated or linked to that other document.

As mentioned earlier, there are other types of deliverables that an architect might produce. Each framework and each methodology has its own set of deliverables. Beyond that, there may be customized versions for each of these documents based on tailored frameworks or a customized or tailored methodology in use in certain companies. An architect must adapt and understand how each of these deliverables maps to the common items described here. Luckily, it is easy to find templates and examples proposed by different architecture frameworks and groups online [8–10]. At the end of the day, remember that the final outcome of the project/architecture initiative is not within the documentation that you produce, but rather with the successful implementation of the project.

8.2.4 Business Technology Roadmap

Another important document or PowerPoint deck, usually provided by an enterprise architect, is the technology roadmap or technology target state. It is usually put together for a group of related technologies (such as middleware, collaboration, messaging and integration, Web and portal, and so on) and contains:

- The scope of the presentation/document: Usually the scope has multiple dimensions, such as the lines of business included (as this might not apply to the entire enterprise), the technologies, or the services in scope.
- The business strategies and drivers that influenced the roadmap/target state such as:
  - Cost management: Increasing the level of automated business processes and using modular designs to support demand-based growth.
  - Business agility and time to market: Using integration capabilities as a service. Increasing standardization by eliminating customized adaptors required to provide required transformations and leveraging enterprise integration services instead.
  - Risk mitigation: Conducting independent audits to review overall facilities and security and compare against industry best practices.
- Business scenarios: What are the various scenarios, from a business point of view, where this group of technologies will be used?
- Current state assessment: Define the current state of these technologies for the various lines of business. What are the current problems and

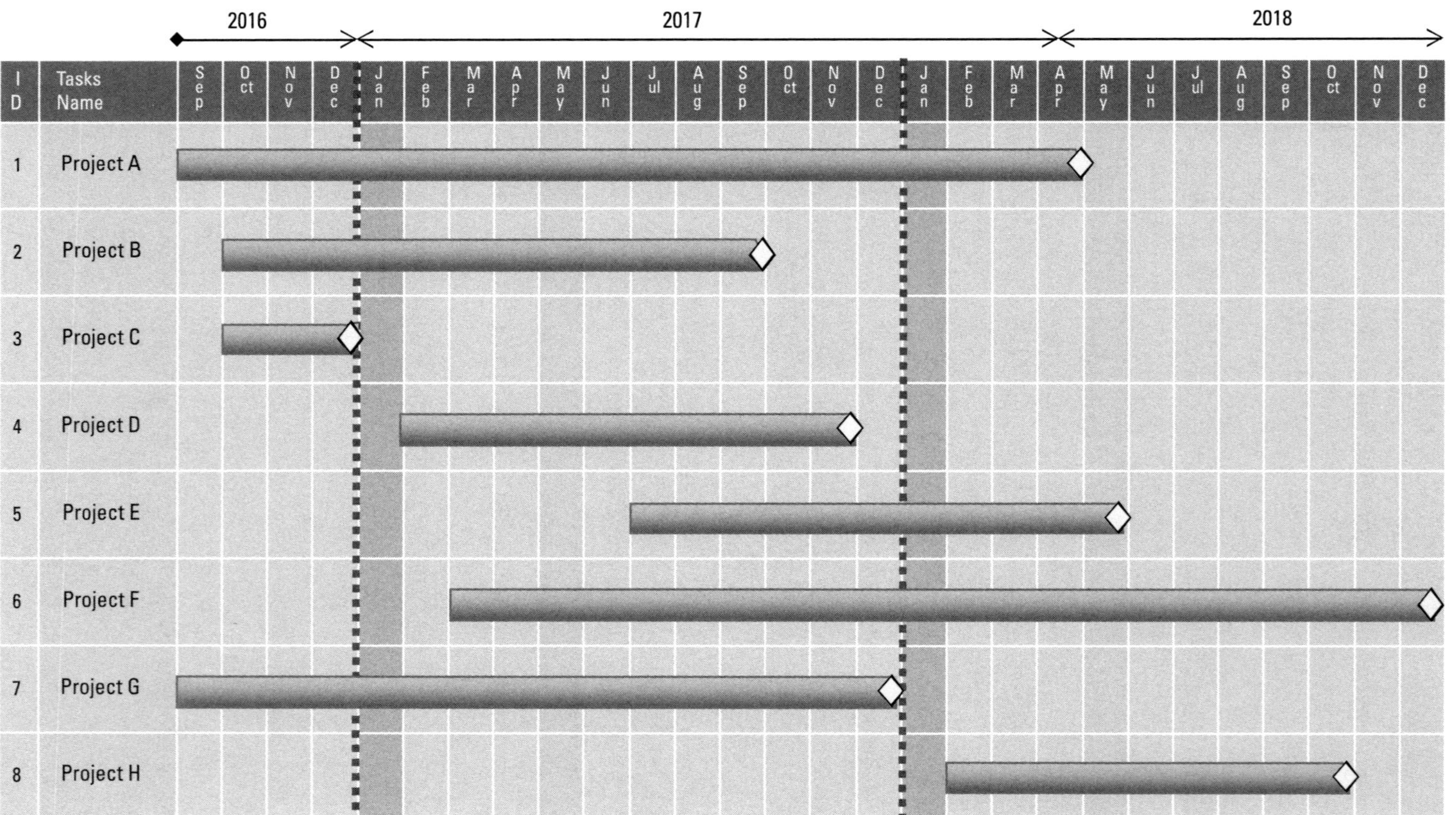

Figure 8.4　Example of transition steps and events.

what are the implications? Some examples include those shown in Table 8.5.

- Target state objectives and directions: These are objectives based on previous enterprise initiatives that should be reached through use of these technologies. There should be a short description for each objective, as well as options for reaching that objective.

- Target state recommendation: Discuss the recommended and eventual interim target states for each of the technologies presented. This should include a discussion for each of the business scenarios identified in the previous section. The patterns, reference models, or reference architectures that have been previously identified and that have influenced the target state should be mentioned in this section.

- Transition steps and events: This section should identify the various work packages/projects/programs that would ensure the transition from the current state to the target state. This might include different implementation milestones for the target state for each of these technologies (please refer to to Table 8.4).

Now that we have seen the input into the architectural deliverables as well as the most important documents produced by different IT architect roles, we will continue our journey by discovering the benefits of the architecture

Table 8.5
New Components

Problem	Implication
Lack of AAA capability maturity	Limited in-house experience with COTS horizontal integration technologies and integration standards used by application and services vendors. Lack of strategic focus on application integration and services-oriented architecture. Lack of enterprise services to integrate delivery of services to customer, or to support inter-LOB application integration.
Tightly coupled applications stemming from point-to-point integration	Lack of agility and high cost of application change caused by: Brittle and inflexible coupling of applications, lack of foundation of re-usable services for new integration initiatives, limited integration due to the prohibited cost. Point-to-point integration is complex and breeds a web of dependencies that offers no reuse. Mediation and service orchestration capabilities are embedded in the applications. Lack of standard or logic/service reuse focus.

practice and what should be the steps to follow to establish or enhance the architectural practice in your organization.

References

[1]   Lockheed Martin Federal Systems, Odetics Intelligent Transportation Systems Division ITS implementation Strategy example, www.iteris.com/itsarch/documents/imp/imp.pdf, accessed May 2016.

[2]   Lockheed Martin Federal Systems, Odetics Intelligent Transportation Systems Division ITS Implementation Strategy example, www.iteris.com/itsarch/documents/imp/imp.pdf, accessed May 2016.

[3]   Functional Requirements document, http://doit.maryland.gov/sdlc/documents/func_req_doc.doc, accessed May 2016.

[4]   Johnson, E., "Good Sample Business Principles," http://ericjacobsononmanagement.blogspot.ca/2010/09/good-sample-business-principles.html, accessed May 2016.

[5]   Define business principles and policies, http://www.iccwbo.org/Products-and-Services/Trade-facilitation/9-steps-to-responsible-business-conduct/Nine-steps/Steps/Step-4-Define-business-principles-and-policies/, accessed May 2016.

[6]   Nygard, M., "Architecture Decisions Examples from Institute for Software, Think Relevance," blog describing the architecture decisions for Agile projects, http://thinkrelevance.com/blog/2011/11/15/documenting-architecture-decisions, accessed May 2016.

[7]   Tyree, J., and A. Akerman, Capital One Financial, "Architecture Decision Template," https://www.utdallas.edu/~chung/SA/zz-Impreso-architecture_decisions-tyree-05.pdf, accessed May 2016.

[8]   Generic TOGAF template for an implementation and migration plan, http://www.penos.co.nz/tgogaf/TOGAF_9_Templates/Deliverables/Phase%20E/Implementation%20and%Migration%20Plan/TOGAF%20%20Template%20-%20Implementation%20and%20Migration%20Plan.doc

[9]   Shupe, C., and R. Behling, "Developing and Implementing a Strategy for Technology Deployment," www.arma.org/bookstore/files/Shupe_Behling1.pdf, accessed May 2016,

[10]   Lockheed Martin Federal Systems, Odetics Intelligent Transportation Systems Division ITS implementation Strategy example, www.iteris.com/itsarch/documents/imp/imp.pdf, accessed May 2016.

9

How Does Architecture Bring Value to an Organization?

So far, this book has covered the reasons to become an architect, the background needed for the job, the skill set, the education, and the deliverables that the architect might produce. This chapter now shifts focus from the individual to the organization. The question every employer asks is: "Why should we spend money from our budget to hire an architect?" Previous chapters observed that the cost of an architect is high compared to other positions that may offer the same services. Thus, it can be hard to justify to a manager why a new resource is needed, let alone an expensive architect. Looking at this from the perspective of the person to whom the architect would report, previous chapters have discussed various types of architecture and the reasons why a company needs an architect. This chapter focuses on reasons that are common to all types of architecture. Although architecture may be viewed as the "new kid on the block" and architects as "people who don't do much and get a good salary," most successful companies did not implement architecture into their organizations because it was fashionable. They did so because they realized the benefits outweighed the costs and because they were able to tie business results to changes proposed, and ultimately delivered, by the new architecture group.

9.1 Benefits of Architecture Practice

There is widespread agreement that any serious company should question why they need information technology (IT) architecture. The phrase "Show me the money" from the Tom Cruise movie, Jerry Maguire, comes to mind because this should be the mantra that all chief executive officers (CEOs) and chief information officers (CIOs) use daily to boost profits and increase shareholder value. The question here is whether the benefits that the architecture brings to the organization outweigh the costs of hiring architects.

Architecture was not always a recognized discipline included in software and infrastructure development. Nor did architecture always play into the decisions on how a group could work on projects to achieve expected results. When software development first started, for example, testing was also not recognized as a separate discipline and most of the time the project would either fail due to the sheer volume of defects, or the final product would not be used as it failed to satisfy the client's requirements. Little by little, this started to change. Systems, applications, and organizations evolved; people started to understand that they needed a way to bring order to this chaos. The lines of business for each organization were functioning almost like separate companies with little or no way to leverage common functionality. There was little or no integration between systems and applications and, as a result, the cost of maintaining them was high.

Because of this fragmentation it was difficult to decide what changes had to be applied at each level when company strategy changed. One by one, companies began to understand that they needed a role that assumed high-level design responsibilities for all systems, applications, and infrastructure as well as for the enterprise as a whole. The architect role was formalized, although its definition was not yet clear. Some people wrongfully assumed that an architect should be the ultimate specialist in each application or system, while others understood that this role actually envisions and manages high level design at the application, system, and infrastructure or enterprise level (please refer to Figure 9.1).

Soon the benefits of having an architect and (correctly) using his or her expertise became clear:

- Applications could be designed to implement functionality required across multiple lines of business, thereby increasing reusability and decreasing the cost of such implementations. In some instances, this led to huge cost savings, especially when the same business capabilities or processes had previously been implemented by different systems across multiple lines of business.

- The infrastructure could also be designed to serve the company as a whole reusing various devices (firewalls, load balancers, UNIX boxes, storage) across multiple lines of business. With the help of an architect, the enterprise cloud constructs a clear plan for the introduction of new technologies, correlating them with projects that were responsible for delivering them.

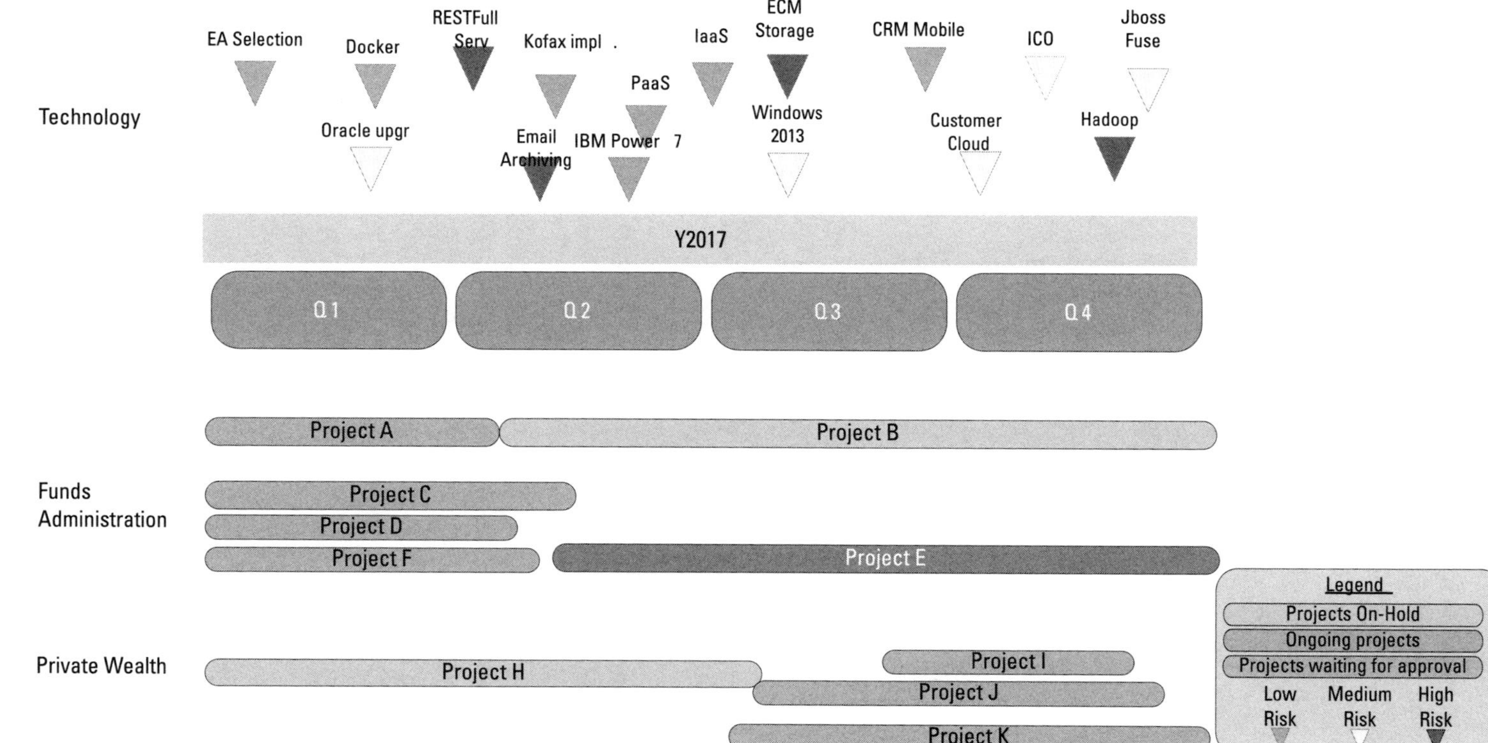

Figure 9.1 Example of technology deliverable showing the technologies introduced and projects to deliver by line of business.

- The security of the IT systems could be centralized and validated by somebody who understood what might disrupt the business, what risks might be involved, and what loopholes had to be closed.
- The list of projects implemented would be driven by enterprise strategy so the CEO could rest assured that financing them would push the company in the right direction toward established goals.

Almost no one would now argue that architecture is useless, but the *resistance to architecture* does not always manifest clearly or directly. On paper, architecture sounds great and serves the interests of the company, but some business or IT leaders might fear losing power. Because some applications are common across the company or even across multiple companies, they no longer have full control of their functionality and these applications and systems do not belong to the specific line of business.

Introducing architecture into an organization is a big game changer, and any kind of change has to be properly managed. Thus, for example, proper management could include identifying the resistance leaders and working with them one by one to make sure they understand why this change is necessary and how it will better serve the ultimate goal of business growth. Enterprise architecture is about building relationships among business units and revitalizing processes and capabilities to make companies more flexible and more efficient in managing change. Some business executives see the business benefits of architecture as risky and elusive compared with the cost of implementing it. The best way to overcome this resistance is to show stakeholders the reasons why imposing consistent architecture at various levels within the organization is important and valuable:

- The term "enterprise architecture" is a buzzword that is not well understood. Just like the latest version of a smartphone, everyone wants to have it, but very few understand why or the benefits of using it wisely and well.

- The company has to respond to more changes from a legal point of view and to come up with new products faster to lead or catch up with competitors. For the insurance industry, for example, this might entail the ability to launch new insurance products to niche customer segments, test responses, and decide whether products should be pulled back, adjusted, or marketed to a wider segment.

- Providing personalized products to customers often requires an ability to combine services that service business functions under the hood. Making such services part of a proper architecture allows the company to easily change the way that they are combined to provide new functionality. Designing applications in silos makes it more difficult to integrate functionality. Sometimes it is easier to redesign the architecture around functionality rather than finding workarounds to combine functionality exposed by existing applications. Architects excel at making such decisions.

- Information is vital to the organization. Taking information out of traditional silos and integrating it into a cohesive whole is one of the main objectives for enterprise architecture. This permits organizations to respond more quickly to changes. An architect needs to focus on how the business can maintain a competitive advantage using current products or creating reconfigurable product architectures.

- Exchanging information between various companies within the same industry or even across industries makes it mandatory to provide standardized service interfaces. These should be modeled and eventually changed or customized to optimize them for the benefit of the company as well as for the entire industry. To understand the complexity of all these interfaces, you must have architecture functions at various levels starting from the enterprise level (to decide how the business goals and drivers will drive the various domain architectures) and going down to the business, application, and technology levels.

- Seeking competitive advantage is part of the survival kit and is not a luxury tool for the leading companies in the industry. Executives shifted their focus from growth to achieving a competitive advantage knowing that they could not get one without the other. In the current competitive environment companies seek new opportunities to gain competitive advantage. This is possible only if the company has an architecture team that understands the architectural landscape at various levels so that they can propose integration of the services provided across different business units and how to improve coordination of business unit strategies with shared resource commitments.

For IT leaders, architecture should be tied to the need to provide the right resource so that systems are designed properly. Best case, such designs take into consideration nonfunctional requirements as well (including performance, security, maintainability, and availability). Both the executive and IT management leaders need to show the benefits of having an architect, as he or she is not a cheap resource. The executive staff usually hires enterprise architects who will produce roadmaps linking the business with IT and the IT management will hire application, solution, and infrastructure architects who will have to take the high-level roadmaps and implement them. As I mentioned in Section 6.1, the most common mistake that the organizations make during this phase (when they realize that they need an architect) is to promote people who are in charge of building the infrastructure and applications. Just because somebody can code, install, or configure systems does not necessarily mean they will be able to design an architecture to enable those applications and systems to interact with each other at a high level. Next, let us drill down and see the benefits for each type of architecture and why each one is important for a business to have.

9.1.1　Enterprise Architecture

Enterprise architecture deals with putting all the pieces together within that big puzzle called the enterprise. Starting from business goals and strategies, enterprise architecture is the wrapper for all the various domain architectures and has as its primary objective translating overall goals and strategies into smaller chunks (work packages) that accommodate corresponding changes within various domain architectures.

What is it about enterprise architecture that makes it so important for a company? If it is so essential, why does it need to be sold all the time? Indeed, the benefits of enterprise architecture should be sold rather than architecture itself. It is all about the positive outcomes or effects that enterprise architecture has on the enterprise. The outcomes and their effect on business development should be measurable and it should be obvious how business strategies and goals are actually achieved.

Enterprise architecture is not about gathering all the details from a business and IT perspective. This is only the beginning because the foundation needs to be laid and the information required needs to be provided for the executives to make the decisions that will put the company ahead of its competitors. As long as they see enterprise architecture as a way to help them achieve their business strategy using a well-defined and time-tested framework, they will fully support the enterprise architecture practice as a money-maker rather than a money-taker.

The enterprise needs enterprise architecture not only for compliance purposes (Clinger-Cohen and Sarbanes-Oxley) but to achieve a competitive advantage by:

- Facilitating the translation of corporate strategies to daily operations: What does "faster time to market" mean or "streamline the core process"? Executives can translate these phrases to projects in the old-fashioned way, but enterprise architecture provides a much better path from business goals, objectives, and corporate strategies to the work packages that will implement them, usually by using a standard framework and/or methodology. This translation is one of the major responsibilities of the enterprise architect who maps both the business and IT strategies to the business capabilities and then to the systems/applications. For example, a mutual fund company might have *investment management, statements preparation,* or *client on-boarding* as level-1 business capabilities. The architect has to build a heat map by understanding the state of each of these capabilities, considering dimensions such as people, processes, technology, and so on. He or she also needs to understand the pain points, issues with the capabilities and how the business owner wants to evolve them. In our example, the *fees management* level-2 capability has to be improved to meet the *enhance core business* business strategy or *client on-boarding* needs to be enhanced to ensure a top-notch client experience (please refer to Figure 9.2).

- Providing a holistic view of all the domain architectures starting from the top (business) all the way to the bottom (technology): This facilitates traceability between the business, application, data, and technology layers across all the lines of business (LOB) removing horizontal (between lines of business) as well as vertical (between the architectural layers) barriers. A simple tool like business capability modeling can increase collaboration between the business and IT. It can also provide the basis for major executive decisions such as considering what current or future state capabilities are of strategic importance and should consume the most budget. If a project cannot be mapped to either one of the strategic capabilities or contribute towards achieving future states, then it

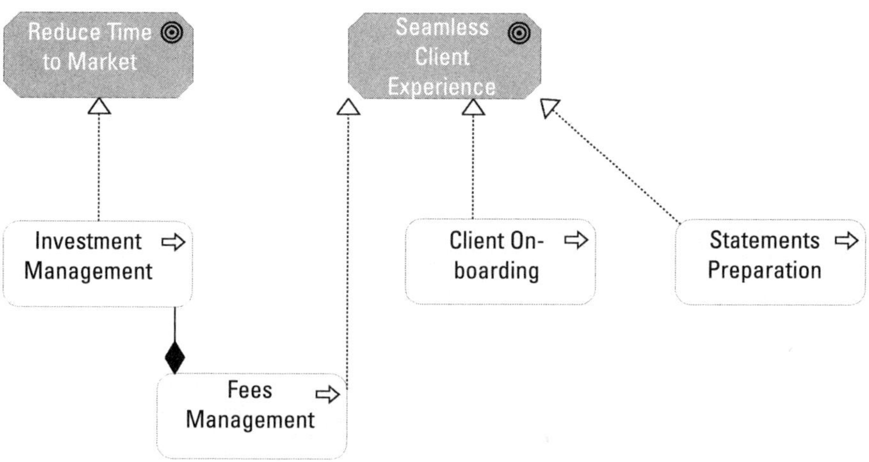

Figure 9.2 Strategies mapped to business capabilities.

should be examined closely and its value should be questioned closely. This picture might also help to decide which business capabilities can be externalized or outsourced so that the organization can focus on core capabilities that put it ahead of the competition.

The example in Figure 9.3 shows how an architect can represent major roles and stakeholders, the business services that they use, the business processes and business functions that deliver those services, and the technology that supports them. The client stakeholder plays the role of the investor who either buys or sells mutual funds. The *administer mutual funds* business function has a flow of business processes and delivers the *buy mutual fund* business service. These processes are automated by the *customer administration service* application service that the CRM System delivers (it might be composed of a series of applications) and is deployed on the WebSphere Application Server sitting on an

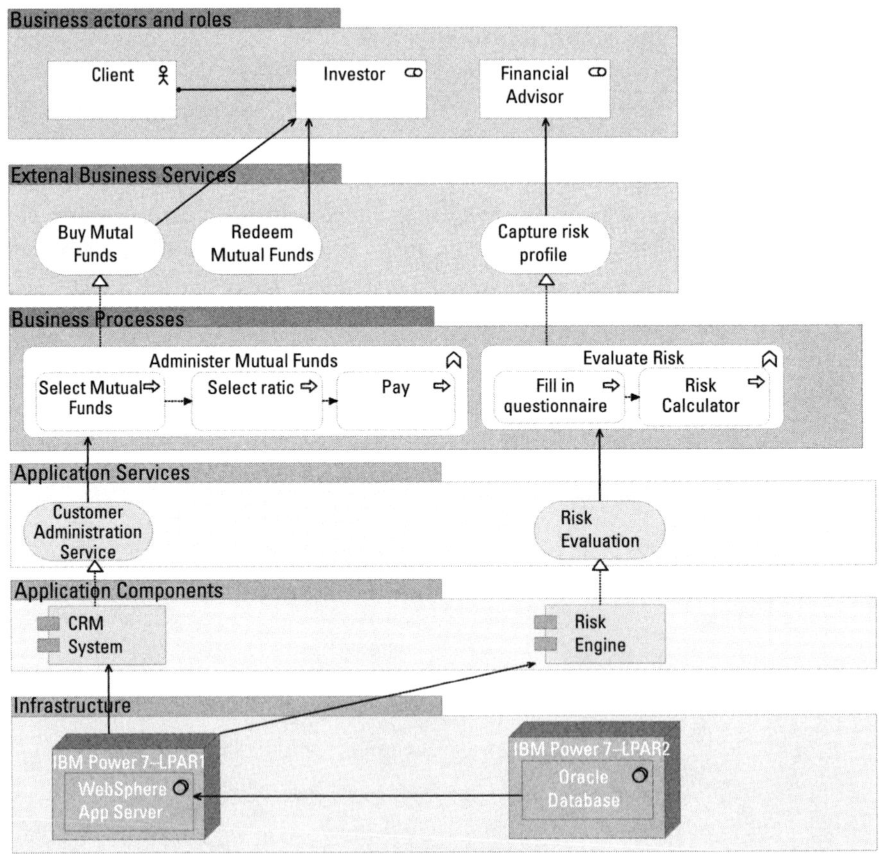

Figure 9.3　End-to-end top-down architecture example.

IBM Power 7 LPAR. For each of these layers, the architect can assess its state to create a heat map for each of the various architecture domains. Archimate can model the architectural concepts, but various modeling languages may be chosen based on enterprise standards, systems already in use, or other requirements.

9.1.2　Business Architecture

The days when the business of a company was simple enough that it could be understood by a few people are long gone. For medium to large organizations especially, the business capabilities, processes, functions, services and roles are becoming more complex with intricate dependencies between them. More organizations cannot afford to run a business that has no documented capabilities and for which the business functions, processes, financials, and organizational structure are not fully optimized. Even if they have a defined business strategy, many organizations do not execute it and do not have a clear link between the strategy and the business capabilities, functions, and processes that have to change to realize the strategy. Recent surveys indicate that there is a direct connection between defining a viable, planned business strategy and achieving intended business outcomes (emergent business strategies are based on day-to-day decisions that set the direction).

As with other types of architecture, if there are no formal methodologies, processes, deliverables, or people assigned to produce them, business architecture will be informally carried out by various roles in the organization, often with poor outcomes. Some organizations have created a business architecture department without a clear direction on its mandate to comply with various regulations. The major benefit of business architecture is to focus on the business outcomes and support business transformation by:

- Defining a solid business strategy that contains guiding policies, diagnosis of the current state, defined target outcomes, and a set of coherent actions to bridge the gap: This path begins with defining business strategies as a set of strategic statements that guide the way to success by defining how a company can achieve its goals, meet customer expectations, and sustain competitive advantage in the marketplace. Some examples of such business strategies include:
  - Put clients first; ensure the best client experience.
  - Diversify and innovate by changing the current products and creating new ones leveraging new technology advancements.
  - Leverage the company's competitive advantage abroad.
  - Invest in employees and create a culture of appreciation to attract best talents.

Once business strategies are defined (and refined), the architect is the person who will take them as input, assess the current state of business capabilities, produce business capability heat maps and define roadmaps to refine the steps required to move from the current to the target state.

Figure 9.4 depicts an example of level-1 business capabilities from the banking industry. The shades of grey of each of the business capabilities was determined by assessing the various dimensions of the capabilities such as processes, functions, technology, information, and people. These business capabilities are further categorized based on the impact on the customer (customer satisfaction and loyalty) and the financial impact (revenue and cost):

- Allowing the business to take ownership and drive the transformation through business centric roadmaps and business models. These models should focus on how the business capabilities will evolve from the cur-

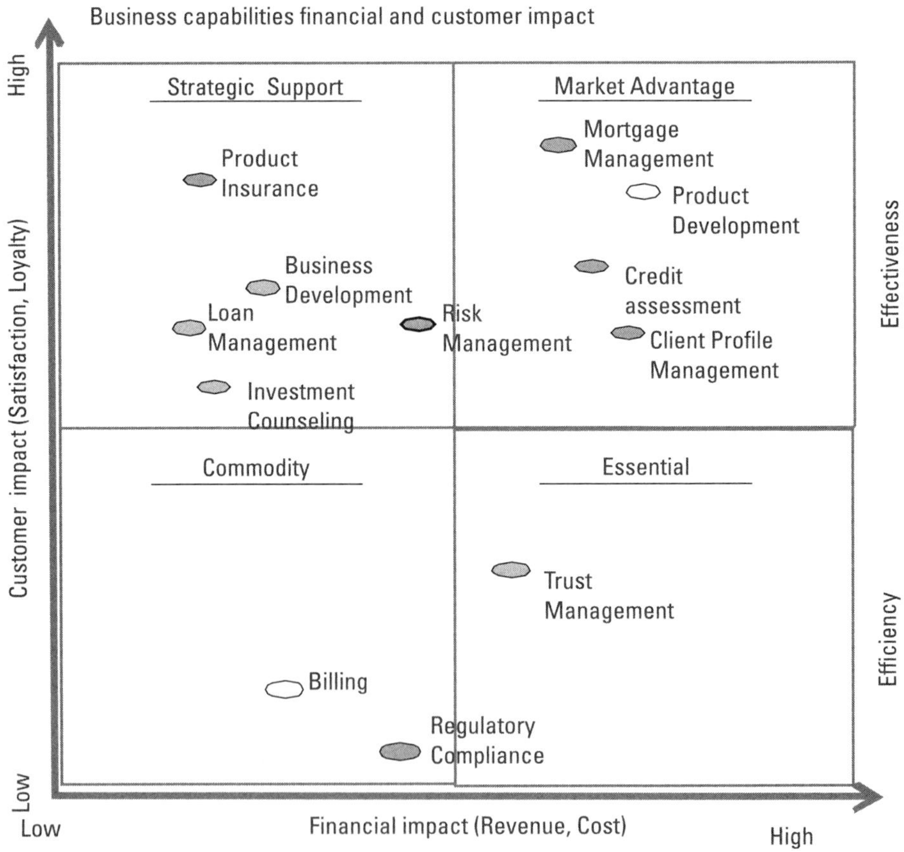

Figure 9.4 Example of business capabilities classification based on the financial and customer impact.

rent to the target state and what initiatives will develop the underlying systems and applications. As an example, Figure 9.4 will become core to the business showing capabilities that the business must evolve to achieve its business goals. An architect can provide further details on how the business capabilities in the *market advantage* or *strategic support* quadrants can be evolved by enhancing any of the underlying dimensions (processes, people, technology, information). He or she can provide this information to the enterprise or solution architects who will produce roadmaps that focus on how business capabilities can be evolved by the improvement of processes, evolution of applications, information, infrastructure or providing training to support staff.

- Bridging the gap between strategy and execution and integrating business and information technology (IT) efforts by defining the risk, implications, constraints, and dependencies of the actions. IT should be an enabler of business strategy with a clear connection between its initiatives and business success.

- Showing the link between operational and strategic views. Business architecture is a useful tool that ties strategic views with operations cost savings and helps clarify the impact of business decisions on the company's ability to successfully operate by supporting the target-state vision.

- Creating a clear prioritization of initiatives for investment. Once business capabilities have been documented the cost of the initiatives can be overlaid, the technical aspects can be translated into business capabilities enhancements and provide the basis for the analysis and business decisions for management. This matrix may have multiple dimensions to help illustrate such different views as:
  - How investments would impact different type of the customers;
  - What business processes support the capabilities and how they can be optimized;
  - Which business capabilities focus on innovation, which provide differentiation (from the competition), and which are commodity-based.

- Defining the core business capabilities that differentiate an organization from its competition. Create a 2-by-2 matrix with the business value and the quality/condition of the application as its axes. The quadrants will determine if the organization should continue to invest in a capability, if it should outsource or migrate that capability to the cloud, or if it should eliminate it entirely (refer to Figure 9.5 for an illustration).

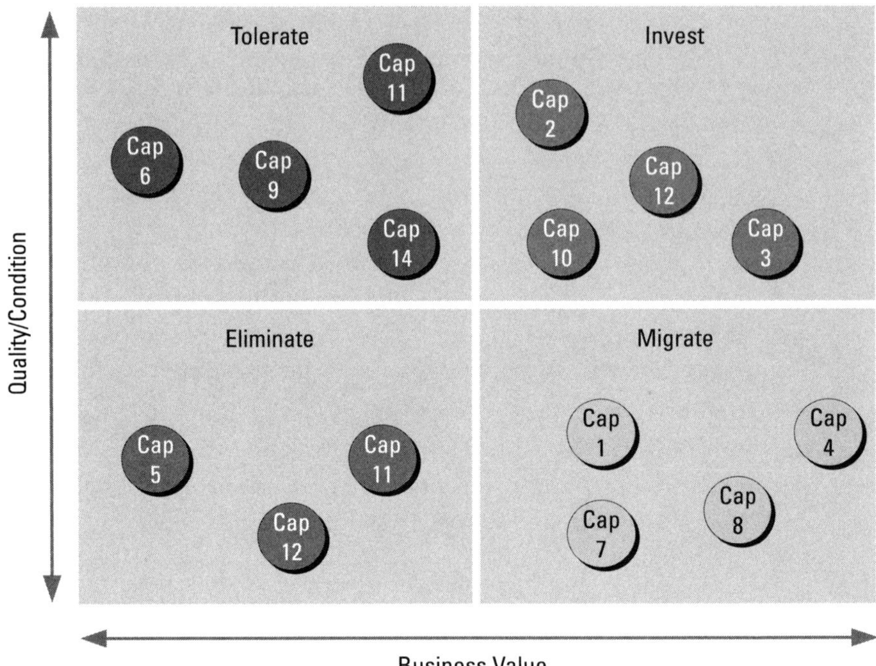

Figure 9.5 Business capability matrix.

- Increasing the reusability of underlying business processes by identifying those that work in common among different business capabilities and eliminating redundant processes. Business processes may be delivered by application services and components, so such analysis can go even deeper and reveal redundancies at lower architectural levels (application, data, and infrastructure).

- Improving cross-functional (or line of business) planning and budget preparation and ensuring that funded initiatives do not work at cross-purposes. Make sure that money is spent on evolving business capabilities will help the organization; turning the most profit provides an important benefit.

9.1.3 Application Architecture

As applications became more complex, organizations felt the need to lay a solid foundation. They sought to document the blueprints for these applications and how they interact with each other to provide business functionality. Such applications must be based on specific requirements and business scenarios, to be sure. The application architect is the person responsible for putting together the documentation that shows the components that make up these applications and how they interact with one another through interfaces. This might sound a lot like a technical (or detailed) design, but such a design usually operates at multiple levels where the application architecture is the highest level only from the application point of view.

Developers often jump directly to coding, deferring design to a later stage, perhaps operating under the impression that they understand fully what the application does and the business logic it needs to implement. For small projects this is usually true, and even for large ones it is a case of producing just enough documentation, avoiding anything that does not bring either immediate value to the project or long-term documentation of the overall system architecture. Application architecture is usually the best place to document any of the architecture decisions that affect the high-level design of the application as well as the technical risks (again on the application side) and how they were mitigated. None of these things would go into a detail design document, which is mostly concerned with the implementation of the various application components. Jumping straight to the detail design (or even worse to coding) could make the final software unstable and difficult to maintain or could even prevent the application from working when deployed into a production environment.

Application architecture acts as a link between business requirements and actual implementation because it addresses all the business scenarios and various paths documented in the use cases. Its main goal is to incorporate the use cases and provide a solution showing how the application will satisfy their requirements. Existing tools provide complete traceability from requirements, analysis, and design models through to implementation and deployment, but as the application architect, you need to make sure that each business scenario is addressed by (parts of) the architecture solution to be implemented by the application. Documenting the application architecture gives stakeholders and the project team confidence that all the major technical issues are addressed prior to implementation so that there are no surprises later on in the project.

Application architecture uses business architecture (either formally defined in the process of establishing the enterprise architecture or informally in e-mails or meeting notes), technical requirements, and directions with regard to the various technologies, patterns, and frameworks that should be used to conform with similar applications already existing in the enterprise. This is exactly the information that a developer lacks when he or she puts together a detail design or starts coding an application.

Another important factor that needs to be considered when producing an architecture is the trade-off between nonfunctional (or quality) requirements. Performance versus security is the most common example because you cannot get a truly secure application without affecting performance and, vice versa, you cannot have an application that performs very well without compromising security features like authentication and authorization. Even the most experienced developer cannot focus on coding details as well as on the multitude of nonfunctional requirements that have to be considered, which might also include:

- Maintainability of the application as a whole: Does a certain development framework that is well understood in the organization need to be adopted so that maintainability is not reduced?

- Reliability: How can the application be designed to be resistant to failure? Thinking about how the Web site cannot only provide the required functionality but also can be reliable even at peak hours, and deploying it on a cluster of application services can make the difference between business as usual and loss of business as the Web site fails and cannot handle any more requests.

- Privacy: How to ensure the client's personal data will be stored and used appropriately. Ensuring the protection of client information is paramount to most companies because scandals (such as the leak of the Panama Papers) have demonstrated that a lot of companies do not take information security as seriously as they should.

- Scalability: How to ensure that the application is able to handle future increases in workload. For example, an application architect can ensure the application will be able to process large amounts of data at peak times by including fault tolerance in the design and composing the database layer from multiple servers organized in a cluster (e.g., Oracle RAC).

- Interoperability: How to ensure that the application will easily interact with future applications. The application architect can design the application to use micro-services that can be deployed anywhere and use Docker as the micro-container, therefore ensuring the ability to implement the services using various technologies.

- Communication: How to make sure the most appropriate protocols are used.

Without a high-level blueprint of the application, there is no confidence that the code implemented will actually perform the functions required by the business. There will be certain times when the architecture will have change. If the architecture is flexible enough to accommodate such changes, parts of the architecture that represent the greatest risk are identified in the plan and the assumptions and constraints are validated and factored in. That is a much better position to take than simply scrambling through the code to identify what needs to change.

From a technical perspective, defining and implementing nonfunctional (engineering) requirements can make the difference between a good application or system and a bad one. The purpose of the architecture is to also to mediate trade-offs between the various engineering properties. In an ideal situation, the architecture would satisfy all the functional and engineering requirements, but in real life there will be a few trade-offs among the various engineering requirements. As an example, an application that requires high performance may be tied to a particular platform and hence not be easily portable. The architect is also the watchdog who makes sure the number of technologies, patterns, or frameworks supported is limited, because otherwise maintainability will be reduced. The architect is the key person to make explicit to the stakeholders all the compromises that the application requires, to prioritize the nonfunctional requirements and to document this in the architecture design decisions. Part of the difficultly is that quality attributes are not always explicitly stated in the requirements or adequately captured by the requirements engineering team. That is why an architect must be associated with the requirements gathering exercise for the system, so that they can ask the right questions to expose and nail down the quality attributes that must be addressed. Understanding the quality attribute requirements is merely a necessary prerequisite to designing a solution to satisfy them. Conflicting quality attributes are a reality and creating solutions that that adequately satisfy these requirements can be quite challenging.

As with other architectural types, at the end of the day, the value of the application architecture is demonstrated by providing some key performance indicators and then reporting on them. These indicators have to be particularly geared towards showing how the architecture actually made the applications more maintainable because it:

- Addressed the foundational (core) parts of the application;
- Provided a flexible design for parts that are likely to change;
- Documented key assumptions and constraints imposed on the solution;
- Documented key technical risks and how the application will mitigate them.

In summary, the application architect plays an essential role in delivering good applications that can be easily maintained and altered in keeping with changes to the business architecture or requirements. The CIO or the sponsor of the architecture in the organization has to understand what the impact over the applications developed will be and how this will save money and provide a sound application backbone in the long term. The application architect has to be able to sell the architectural solution and gradually steer the enterprise toward a successful realization and implementation of their business goals. This translates into steering the architecture towards strategic solutions while also maintaining a balance between strategic and tactical goals. If, for example, a new industry regulation requires the change of an application that relies on a sunset technology (that will be retired soon), then a deviation (waiver) can be obtained to implement this change on the legacy system, thereby adopting the tactical solution. If new application services need to be developed, the architect will sell the strategic solution and influence the stakeholders to implement these services on the new platform.

9.2   Establishing the Practice

Once all these benefits are clear and you start running with the mandate to make it happen, the task is not easy. As with any change, there are many possible implementations and whichever one is picked must deliver the benefits mentioned earlier.

Some of the architecture frameworks (TOGAF, for example) recommend using the architecture framework itself to establish an architecture practice, as this is a complex endeavor and you want to make sure that you get it right the first time. At a minimum, the sponsor of this change [CEO, CIO, vice president (VP), or (IT)] needs to engage a team to:

- Identify the stakeholders and their concerns. What do they expect from the architecture, are these expectations valid, can you achieve the target state (having the architecture practice implemented) in the timeline proposed?

- Identify the main drivers for this change and document them. The best starting point for this is by brainstorming the pain points for the business and the IT managers' experience of issues on a daily basis.

- Define the scope. Make sure it is clear what problem needs to be resolved. The terms architect and architecture have many meanings in the IT, world so you want to be very clear what kind of architecture you are trying to implement. It may be application architecture to ensure the overall high-level design and application programming interfaces (APIs) between applications are documented, it may be business architecture to document the business processes, functions, roles, actors, and data, or it may be enterprise architecture, documenting how the business strategies, drivers, and constraints would translate into projects. Depending on the size of the company, there might be different types of architecture. A medium to large bank, for example, with thousands of systems would require almost every type of architect discussed so far: application architects associated with development groups, solution architects to ensure the alignment between the enterprise direction and the actual implementation, infrastructure architects to ensure that the infrastructure technologies are a good fit for the needs of the company, and so forth.

- Define the architecture accountability matrix. These are the roles and the accountability of each role for architecture deliverables and processes. This matrix should also include required architecture governance structures and roles. The structure can be designed and implemented in iterations starting with the architect roles most needed for the enterprise and then adding other roles later on. As an example, at a minimum, an enterprise can have:

  - From a process point of view: Defined gates or checkpoints such as after the requirements or conceptual architecture phase, after the architecture or design phase, implementation phase, and so on.
  - From a personnel point of view: Solution architects working as project architects reporting into lead architects having responsibilities for each line of business, and an enterprise architect to define the roadmaps based on business goals. Later on, other architect roles can be added such as application, data, infrastructure architect, and more.
  - From a governance point of view: Define an architecture review board composed of lead architects and senior management to revise only the complex initiatives. Later on, other architecture boards and checkpoints can be added.

- Define the architecture performance matrix. The metrics that should be used to monitor the performance of the architecture practice against the stated vision and objectives.

- Define the architectural governance and how it ties into technology, IT, and corporate governance. This is related to management and control for all aspects of the development and evolution of the architectures. It includes the processes required to identify, manage, audit, and disseminate all information related to the implementation of the architecture, as well as its organizational structure (architecture review board, compliance team, and so forth).

As with the other initiatives, establishing an architecture capability can be an initiative by itself composed of multiple work packages like establishing the architecture team, coming up with the architecture processes and governance, and so on. Once the architectural practice is implemented, any changes or additions to the various governing bodies should be assessed and implemented following standard methods and best practices.

9.2.1 Practice Owner

There should be an owner who will be responsible for setting the direction for the practice and maximizing the value the architects will provide to the company. Even before starting to define an architecture practice within the organization, a practice owner should be selected or hired with a clear understanding of what is the mandate will be for his or her team (what is the target state?). Like an architect, a process owner must do the following:

- Assess the current state. Are there any formal or informal architecture groups in the organization, what is their maturity, and what kind of services do they provide?

- Assess the target state. What should the architecture group look like at the end of the transition, what services should the group provide, and what skills or competencies will they require to provide the services?

- Come up with a plan on how to bridge the gap (either by hiring new people or by training internal staff) and *sell* this plan to the senior management executives focusing on the benefits of the architecture for the organization. The practice owner has to become an architecture champion and be able to convince senior management of the benefits and value that the architecture will bring. Planning for change involves preparing different presentations to develop awareness within the rest of the organization, as well as preparing a budget for hiring and training staff to perform the required activities.

A practice owner requires not only a thorough understanding of the architecture practice but also an understanding of the other practices within the IT department and what dependencies there are between them and the architecture. Typically, he or she is a former architect who has moved into a different role. Ideally, this person has prior experience in shaping the architecture team at other organizations.

A practice owner will need the support of senior management to establish the practice, but he or she is the one who must define and implement this change. This is not only about the kind of people that should be hired to perform the role or about defining the role itself (creating the role summary). It is also about defining how the architect role will interact with other positions and what processes are related to the architecture practice.

9.2.2 Defining the Practice

Some of the most important activities of the architecture practice include:

- Establishing an architecture board: This is composed of all key stakeholders responsible for review and maintenance of the overall architecture. A few such responsibilities are:
  - Ensuring consistency among architectures;
  - Identifying reusable components;
  - Enforcing architecture compliance;
  - Improving the maturity level of the architecture discipline;
  - Providing the basis for decision-making regarding changes to the architecture;
  - Providing an escalation capability for out-of-bounds decisions;
  - Ensuring consistent management and implementation of the architectures;
  - Providing advice, guidance, and information;
  - Ensuring compliance with all architectures, and granting dispensations in keeping with the overall technology strategy and objectives.

  The TOGAF methodology contains additional information about what could trigger the establishment of an architecture board, what should be its size and structure, and the recommended operation from the governance perspective.

- Architecture compliance: Once an organization has an enterprise architecture roadmap that clearly defines the current and target states for various domain architectures, an architecture compliance process is needed to ensure compliance of the project architectures with the enterprise architecture.

  This process typically includes gates or checkpoints where the architecture presented is scrutinized against established architecture principles, business objectives, and values. This ensures that any errors in architecture are caught early so as to reduce the cost of later change, as well as identifying potentially reusable services and investigating how well best practices have been applied. A typical architecture compliance review process is depicted in Figure 9.6.

- Architecture governance: This is a practice in which each level of architecture is managed and controlled to ensure implementation and evolu-

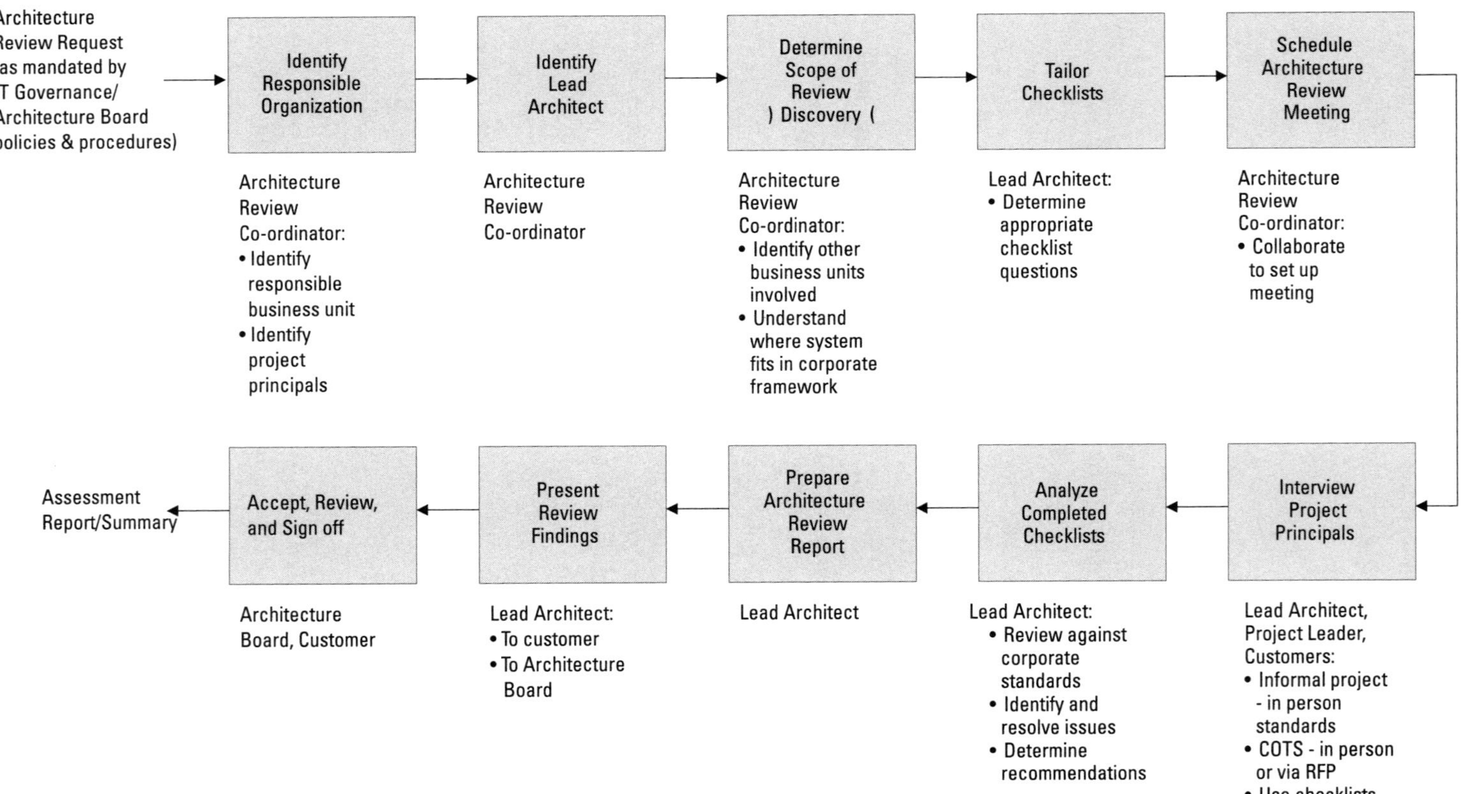

Figure 9.6　Architecture compliance review process. (Source: TOGAF version 9.)

tion of various architectures, compliance with the internal and external standards, best practices, and accountability to clearly identified stakeholders. Architecture governance includes:

- Processes: Policy management, compliance, dispensation, assessment, and selection of the technology and products and compliance with the organization's business policies.
- Content: Regulatory requirements, service level agreements (SLAs) and operational level agreements (OLAs), organization standards, technology/product set, and architectures.
- Repository: The place where the content is usually stored. To establish proper architectural governance across an organization, three elements must be in place:
  - Architectural board: Cross-organizational with the backing of top management;
  - A comprehensive set of architecture principles to guide, inform and support the way the architectures are developed;
  - Architecture compliance: To ensure compliance among architectures.

- Measure the architecture level. Determine the architecture Capability Maturity Model of the organization based on the following elements: architecture process, development, business linkage, senior management involvement, architecture communication, IT security, architecture governance, IT investment, and acquisition strategy. The various levels can be: none, initial, under development, defined, managed, and optimizing. It is important to gauge these levels and then to track them as the evolve and advance.

- Architecture skills. This includes a good starting point for the architectural roles, skills for each of the roles, and the depth of knowledge. Currently, there is a lot of confusion about the various architect roles. Also, the definition or responsibilities for those roles varies from one company and industry to another. This lack of uniformity makes it hard for companies to recruit or assign staff to fill architecture positions. The TOGAF framework has a section (Architecture Skills Framework) that defines a number of roles including architecture sponsor, architecture manager, enterprise architects, various domain architects such as business, data, application, and infrastructure as well as different kind of skills that include enterprise architecture skills, project management skills, IT general knowledge skills, technical IT skills, and legal environment skills. These roles and skills as well as associated proficiency levels (background, awareness, knowledge, and expert) will give the organization a framework for building role profiles for the various architects and to enable them to customize various roles to meet an organization's specific needs. As an example, an enterprise architect role might also include some project management skills or responsibilities related to more in-depth systems architecture or infrastructure architecture.

Hiring the wrong person because the architecture role is not properly defined and understood by the hiring managers will increase the costs owing to the need to rehire and will adversely impact the time, cost, and quality of the systems and the projects that deliver them.

In terms of the complexity of a solution, a certain organization might require different types of architects who would have a specific focus within various phases of the development process:

- Enterprise architect: This person is responsible for architectural design at the enterprise level, producing various business and technology roadmaps, governance artifacts (policies, standards, and catalogs), request for proposals (RFPs), assessment, and reviews.
- System/portfolio architect: This person is responsible for the architectural design and documentation for specific accounts/lines of business within the organization. Their focus is on the enterprise-level business solution within the account.
- Solution architect: This person is responsible for designing the systems or subsystems that function at a lower level of detail than the other architects and focuses on technology solutions.

For each of these roles, the practice owner (usually the manager of the group) will develop role profiles that should contain:

- Various activities that the architect must perform such as providing consulting to various groups, producing the architecture deliverables, and approving architecture waivers;
- Core skills and competencies that each role needs to be able to perform;
- Levels within each type of the architect job such as novice, intermediate, master, and expert;
- A list of behavioral indicators for each of the skills and competencies for each level;
- A matrix to map the activities to the skills and competencies, including the level of expertise required for each of the levels.

9.2.3 How to Hire the Right People

Establishing architecture practice within an organization is no easy task. Once executives and management decide to pursue this path, a first step is to hire a consultant (preferably with an architecture background) to:

- Clearly define roles and responsibilities, tailoring one of the existing frameworks for the needs of the organization. Not every organization needs all types of architects described above. That is because the same person might be able to work as both enterprise architect and portfolio architect. Once a company's accounts or lines of business grow, it may need to redefine those roles and hire more people as portfolio or solution architects.

- Prepare interview questions for each of the architect roles and set expectations that successful candidates must meet. These interview questions should focus on previous experience and should include different sections depending on the role involved:
  - For enterprise architects: Questions about the value of the enterprise architecture, various methodologies and frameworks, and how they should be tailored to bring the most benefit to the enterprise, some typical enterprise architecture deliverables including architecture vision, roadmaps (and what information they contain), as well as how to evaluate the conformance of a certain system/solution architecture with the enterprise architecture.
  - For business architectures: Questions about the methodology or framework used to produce business architecture deliverables including how he or she documented business strategy, vision, business capabilities, processes, policies, and rules that allow an organization to carry out its mandate and functional responsibilities. Questions about how he or she helped to execute on the business strategy, deliver specific business outcomes, and align with IT are also important. Depending on the business architecture methodology used within the company, there might also be questions about value streams, stages, items, relationships between them and the business capabilities, techniques for interviewing the stakeholders, business capabilities, heat maps, and what dimensions were used to produce them.
  - For solution architects: Questions about how the individual worked on documenting the architecture or technical requirements, what architecture framework and methodologies they used, what kind of level of architecture they captured with the diagrams (contextual, conceptual, logical, physical), how they managed the relationship

with the stakeholders and how they presented the solutions to the management, what kind of roadmaps they delivered, and so forth.

- For application/software architects: Questions about the various nonfunctional/engineering properties of an application including performance, scalability, availability (failover, high availability, cluster, and more), maintainability, reliability, recoverability, as well as well-known design patterns, service-oriented architecture (SOA) (concepts, patterns, implementation), as well as different software development methodologies (e.g., RUP, Agile, and more).

- For infrastructure architects: Questions about the infrastructure architecture design patterns, how various designs the candidate created were in line with enterprise architecture including policies, procedures, and how he or she was able to propose and implement best practices and take into consideration various nonfunctional (engineering) requirements such as performance, reliability, and maintainability for the various infrastructure systems involved.

- Modeling (all architect roles): Questions about modeling languages and tools used at various levels including:
  - Enterprise architecture modeling: Modeling languages such as Archimate that can model various architecture layers including business, data, application, and infrastructure plus goals, requirements, gaps, work packages, and more.
  - Business architecture modeling: Using different modeling languages that are either specific to the business layer or contain objects that can be used to model the business layer (apart from other architecture layers). The most popular modeling languages include business process model and notation (BPMN), event-driven process chain, or business layer objects from Archimate.
  - Application/software modeling: Questions should focus on using UML or similar modeling languages to create a model for the application including structural diagrams (class, object, component, deployment) as well as behavioral diagrams (use case, activity, sequence, communication, state machine). Ideally, a candidate would offer examples of how he or she was able to model at various levels (conceptual, analysis, and design) and how those diagrams and views produced catered to their target audience.

Some examples of such questions include those shown in Example 9.1.

Example 9.1: Interview Question: Modeling/Methodologies Skills

Tell us about a past architecture engagement where you were personally responsible for a design and that you're most proud of… Whiteboard the solution at a high level for us, and describe why you feel this was a good example.

Can you share with us your past experiences with formal IT process methodologies? What types of deliverables were you expected to produce?

- Architecture frameworks and methodologies: Questions would focus on how the architect was able to use the most popular frameworks or methodologies to produce deliverables. This would include customizing a framework or methodology to the needs of the enterprise, customizing the content of the deliverables, and helping in setting up or evolving the architecture practice within the organization. The purpose of such questions is to assess how quickly a candidate could become productive in organization's specific environment. Should the organization use a customized in-house framework, questions should be tailored to assess a candidate's abilities to learn quickly and create corresponding deliverables.
- Soft skills (all roles): Last, but not least, questions related to soft skills. One should ask questions to elicit information about soft skills demonstrated by candidates in previous projects or initiatives:
  - *Communication and presentation:* By far one of the most important skills, together with leadership, questions related to communication skills should test how well the candidate can translate between business and technical language to present ideas or roadmaps. Ideally, these would be easily understood by the audience (what views/ techniques the candidate used in previous projects/initiatives). These questions should induce the candidate to discuss real-life situations when he or she had to communicate solution architectures or present them to management. One example is shown in Example 9.2.

Example 9.2: Interview Question: Communication Skills

What types of experience have you had in adjusting your message or approach for your audience? Specifically, tell me about a time when you had to communicate technically complicated material so that it could be clearly understood by a nontechnical person.

- *Facilitation:* Often bundled up with communication skills, facilitation questions test the ability to facilitate gathering information from business or technical subject matter expert (SMEs) or other stakeholders. It is necessary to assess a candidate's ability to build a path and guide the conversation to reach a decision or certain outcome. In many cases, the architect must facilitate a brainstorming discussion to devise the best solution from various alternatives. One example is shown in Example 9.3

Example 9.3: Interview Question: Facilitation Skills

Tell me about when you directed a discussion to build consensus and ensure that a common understanding was reached between all participants (i.e., clients, team, stakeholders).

- *Leadership:* Interview questions testing a candidate's ability to influence other people without relying on authority or position, using appropriate interpersonal styles and methods to guide individuals or groups. These questions should focus on leading various teams (depending on the type of role), maintaining a positive motivation or setting a positive example. As with other performance competencies, phrase questions carefully and require each candidate to walk through a real-life example from his or her experience. A good example of an open-ended behavioral question to assess leadership is shown in Example 9.4.

Example 9.4: Interview Question: Leadership Skills

Tell me about your greatest success leading a team. Please define a situation where you worked with a team to influence the direction of a technical solution.

- *Negotiation skills:* Testing a candidate's ability to gain agreement or acceptance from the other members of the team or other stakeholders, building consensus, and ensuring that a common understanding was reached between the participants.
- *Planning:* Testing the ability to generate plans to achieve completion of work packages identified in the analysis of the gap between the current and target states (at each of the architecture domains).
- *Teamwork:* Questions related to how a candidate got along with other team members and how he or she was able to build good relationships with them.

The consultant should prepare an interview form for each of the architect roles, including guidelines on what questions could be skipped based on the previous answers. The consultant role could also be filled by the manager/director of the architecture team, as he or she should be (or become) the practice owner.

Initially, the architecture practice may include different types of architect, but once the organization grows and the various roles become more standardized, the teams will separate. The most common pattern is to start with the enterprise, system/solution, and sometimes even the application or infrastructure architects all working in the same team.

As with any other practice architecture will evolve over time, becoming more formal both in terms of responsibilities as well as activities performed, and in the composition of the team. Establishing an architecture practice is an ongoing journey wherein targets and objectives might change from year to year based on the needs of the enterprise as well as the evolution of industry frameworks and methodologies. The role of the practice owner is to be able to steer the architecture team in the right direction to maximize the benefits it brings to the organization.

10

How to Get the Job

If you are reading this chapter, then you probably have formed some ideas about what it means to be an architect, what kind of work you are supposed to do, and what education you need to get there. Let us assume that you did your homework and used the guidance provided in this book to embark on the road to become an architect. You read some of the books mentioned in this book's bibliography, you started to apply some of those concepts in your day-to-day work, and you showed your interest to other people in your organization. If you are lucky, you have also indicated to your manager in one-on-one meetings that your long-term goal is to become an architect, leveraging your experience and using the knowledge that you have acquired so far.

As with everything else in life, nothing happens unless you work toward your goals. After all, the transition to an architect's job means changing careers, leaving the safe haven of your previous profession, and navigating tumultuous waters to make for the architect harbor. This book tries to prepare you as best it can for a smooth journey and to become successful in a relatively short time. Here are few approaches you may want to consider as you follow this course:

- *Grow with the company:* Many companies try to grow their size and re-tain the most talented by allowing them to grow either in their current roles (to a senior position) or by moving into different roles that might allow them to continue to use their business knowledge for the benefit of their organization. This means that they might be allowed to work as an architect on a short-term temporary assignment, which is a great way to experience the actual day-to-day work instead of changing career abruptly (and later on finding out that it is not as hoped or wished for).

  If an organization is medium to large and has a mature architecture practice, there might also be a mentor or coach to help make the transition as painless as possible. At the end of this trial period, both you and your manager will assess your work, your perception, and your desire to work as an architect and decide if you should fully transition to the architecture team or if you will return to your old position. No matter what the result, you must keep an open mind and take this as an opportunity to explore a different side of your personality and your ability to fit into a new profession. You are one of the lucky ones who had this opportunity and did not have to leave your job just to find out after a few months that you do not like being an architect and that you cannot go back. When this happens, it is often because you have dismissed or neglected those critical soft skills such as communication, leadership, and facilitation so crucial for success in the architect role.

If your organization is smaller, you might not have an architecture team already in place. Your role will be a little bit harder to define and fill, because this puts you in the situation of needing to *create* yourself as an architect. In this case, you need to make sure that you have enough ammunition, so please refer to Chapter 9 where the benefits that the architecture brings to the organization were emphasized. Be sure to have some examples available about projects that failed or went significantly over budget owing to the lack of architecture. Then explain how the architect role can ensure proper governance at the various levels (enterprise, application, business, and so on). This is a longer and harder path, but as long as you understand the role and how you can make a difference, you can convince management that the organization needs an architect and you are the right fit for the job. As mentioned in Chapter 9, you must "show them the money" and quantify the technical intricacies. You can demonstrate the need for architecture in terms of money saved by avoiding the failure of projects and thinking about the applications or systems design in the long term and at the enterprise scale.

- An even harder path is to leave everything behind and start your new profession with another organization. This should not be the preferred option because you do not yet have the experience to work as a full-time architect. However, this sometimes might be your only option if your company does not have an architecture practice and does not want to even consider establishing one. Even worse, you might be employed as an architect (working as an application architect in a development team, for example) but your title and salary are not in line with what you are doing. If you have been with a company for quite some time and you know almost everyone there, it is hard to leave because you can always find excuses such as "I have a good job and decent pay, who knows what

I would get out there?" The stronger the bond with team members and/ or your manager, the harder it is to leave. Generally, if you make the leap, you will end up with a more fulfilling and better-paid job. You will miss the people but you can start building relationships and, in a little while, you will again know almost everyone at your new company.

- The final approach is the hardest one because it involves working on your education, getting into a training course, and then passing exams to get certified before trying to find a job. This is a less recommended option because you will have to demonstrate that you will be a good fit. Remember that what an employer wants is to make sure that you will become productive as close to your hiring date as possible.

No matter what approach you end up taking, you must assess yourself and be conscious of your current level of skills, knowledge, and experience. You do have to sell yourself, but you do not want to oversell or undersell, especially from the point of view of your experience. This will help you later on when you change jobs, making sure that you apply for the right level. Then you will not end up struggling with a senior role when you are at the beginning of your career or becoming bored and lacking a challenge when you obtain a job that does not match up with your level of knowledge and expertise. It is OK sometimes to accept a more senior role, but make sure that you do not stretch yourself overmuch, as this could end up in you being laid off or moved to an inferior position. If you obtain a job that requires a higher level of technical, communication, or facilitation skills, make sure that you fill those gaps as soon as possible by acquiring the relevant knowledge and expertise.

10.1 Be Sure to Apply for the Right Role

As covered previously, the architect role is not completely formalized. In previous chapters, I tried to make it clear what various types of architect roles are available, so you can pick the one that best matches your skills and experience. You may want to start with a smaller transition (e.g., from senior developer to application architect) and then move later on to system or solution architect roles and perhaps even into enterprise architect vacancies thereafter.

Organizations do not make this easy because they may have their own variations on the roles mentioned in this book, or they may combine two or more of these roles into one of their own devising. Examples include combinations of lead developer or team lead and application architect, infrastructure specialist, and infrastructure architect or even combinations across multiple disciplines such as project manager and architect or business analyst and architect.

If you decide that you want the job, it is important to determine how you can convince them that you have most of the experience needed and that you can cover any gaps quickly. Sometimes an employer is more interested in how good you are as a fit for the team and the organizational culture. They even might decide to hire you if you do not have all the required knowledge or experience as long as you can demonstrate that you know their culture and can quickly adapt and become productive.

Although there have been some improvements in recent years, there is still a lot of confusion regarding architect roles. Most of the time, the job posting contains a variety of skills and competencies that belong entirely to other roles (such as software developer or project manager). Be very careful when applying for such jobs because you might end up doing something different from what you should normally do as an architect. The most common examples for this kind of role conflation include:

- Code review or even hands-on coding requirements posted in the job description for an application architect: Using an application architect for hands-on coding means that either the company is too small to have a person doing only application architecture or that the role is not properly understood by the practice owner. Sometimes this might not be included in the job description, but it might be implied during the interview. If you are at the beginning of your architect career, you might get such a job as it would ensure a smoother transition from software developer. Otherwise, steer clear!

- Project or product management skills required for the architect role: Although the architect is supposed to contribute to the plan and understand constraints, assumptions, and risks, he or she is not supposed to be a project or product manager. There is a lot more to being a project manager (PM) than just producing a project plan, and you might end up doing a lousy PM job, which is understandable as you would not have the skills or the experience. Even worse, once you got the job, you might have to scramble to get the knowledge or be pushed into becoming a project manager entirely.

- Collecting requirements for the project: Although an architect participates in requirements collection and is responsible for technical and architectural requirements, the business or system analyst is the one who has the skills and experience and is supposed to document them. For the organization, it might seem like killing two birds with one stone (get an architect and an analyst for one salary), but this will again result in poor quality of deliverables.

- Management responsibilities: Because the job is a combination of an architect and manager, such positions might be called something like solution architect or lead senior manager. These kinds of jobs include having direct reports but also putting together a budget or supervising projects. Sometimes these jobs are used as stepping stones towards other management jobs and some experienced architects might want to go this path. If so, be aware that your days of doing architecture 100% of the time are over.

- An enterprise architect being required to deliver system or application or infrastructure architecture for various projects: An enterprise architect might have a background as an application or solution architect so he or she might be able to do the job but the organization has to understand that these are different jobs with different deliverables and activities.

Even for the most standardized role before moving ahead and applying, you might want to look at various other factors like:

- *Role evolution:* What is the probability the role will evolve or grow in the organization? Even if the role is not what you really want (for example, you may need to also act as a lead developer or team lead), once the organization grows and the responsibilities will be segregated across multiple individuals you will end up doing what you signed up for. It might take a while, but in the meantime you will learn business knowledge that will make you a valuable asset to the company.

- *Organizational culture:* How much does the organization value individual training and development for its people? You should choose an organization with a culture geared towards people, which gives them opportunities to develop themselves and keep abreast of the newest technologies, patterns, frameworks, and so forth. The alternative, where you might not get any training opportunities for years (sometimes ending up paying from your own pocket for training, books or passing certification exams), should be avoided at all costs. Try to get insider information from friends, connections, or sites that rate employers such as Glassdoor (https://www.glassdoor.ca/Reviews/index.htm).

- *Salary:* In most surveys, salary is not the most important factor in choosing or changing a job. Nevertheless, you want to make sure that the salary is within the range for the type of architect and industry to which your company belongs. Plenty of sites provide a range of salaries once you input the area, type of architect, and industry. Some examples of such sites include:

- Payscale [1];
- Glassdoor [2];
- Various job sites such as indeed.com.

To summarize, to make sure that you apply for the right role, you need to read the job description carefully. Then try to map the responsibilities, qualifications, and skills to what you have read so far in this book. Look for any specifics like application architect with J2EE, .Net experience, different architecture frameworks, and methodologies that are listed either as optional or mandatory (Zachman, TOGAF, and so on), different architecture tools (EA Sparx, for example, became one of the most popular in recent years), or different architecture paradigms like service-oriented architecture (SOA), resource-oriented architecture (ROA), or cloud architectures.

Do not apply for jobs when you lack some of the required experience (examples are systems, applications, and products SAP (systems, applications, and products) software solutions integration, digital, and so forth) or technology expertise (middleware, messaging, portal, and so on). You need to understand what level of experience is required, because you do not want to apply for a senior job when you are just transitioning into the architect role. Some job descriptions are quite detailed, while others are spare. If you decide to apply, make sure you fill in any blanks during the first interviews by asking the right questions. Do not be afraid to ask; it is much better to avoid a job that is not a good fit than it is to take a job and later on realize that you have chosen unwisely.

10.2   How to Get Hired (or Promoted) as an Architect

So far, this book has given much of the information required to make a decision and then to prepare yourself to become an architect. Starting with what it means to be an architect, what kind of architect jobs exist, what background and skills the architect needs, and what deliverables he or she would produce, it has slowly guided you to a point where you have all (or most) of the knowledge required to become an architect and you are actively planning to change your career and become an architect.

Becoming an architect does not necessarily mean that you must leave your job and start looking for a new one. Most of the time, people transition successfully into architect positions within the same organizations. This is a win-win situation for both employer and employee. The most important asset you have when employed with the same company for a long time is business and culture knowledge. Organizations try to retain long-term employees by offering them promotion opportunities because they have demonstrated that they can successfully do the job but also because they possess business knowledge and intelligence. Many organizations have a slogan in some shape or form stating that people are their most important asset. When something is important to you, you usually take care of it. Some organizations try to keep people happy in their jobs either by providing training or promotion opportunities, growing the people in the current role or in a different one. If this describes your organization count yourself lucky. You must work hard, but at the end of the day (or after a few years) your long-term goals will be realized. For the purpose of this book, we will consider that your long-term goal is to become an architect, one of the "big guys" who make the difficult decisions that influence the long term development of the application, infrastructure, business, or the whole enterprise.

If you are not one of the lucky ones and your organization only looks at you as disposable asset that can be replaced at any time, things are more challenging. If you still want to fulfill your dream of becoming an architect, then you need to start thinking about leaving the company and finding a job as an architect somewhere else (hopefully where your skills and loyalty will be more appreciated). I know it is not easy and it is a lot of hassle to move to a different job (especially if you have been in the current one for a few years), but there can be quite a few rewards when you find a better fit.

10.2.1 Show Them What a Great Asset You Are

Regardless of whether you are promoted or apply for a new job elsewhere, you must look at your situation from the employer's point of view. Try to wear their shoes (not literally) and prepare yourself with the answers that the employer needs. Apart from the obvious, being prepared and having the right background, skills, and competencies to do the work, what extra quality, the proverbial "cherry on top," should make them hire you instead of somebody else? All (or most) of the candidates who apply have what they need to do the job, but being picked as a successful candidate might be because:

1. *You have confidence in yourself.* As Theodore Roosevelt said, "Believe you can and you're halfway there." Some people might argue that confidence cannot replace knowledge, experience, and other factors enumerated next, but having confidence goes a long way. Interviewers (usually) do not know you, but seeing how confident you are in your abilities to do the job will put you halfway there. This must be backed by knowledge and experience (and any sign that you might lack them would make you quickly lose the interviewer's trust), but showing that you are confident in yourself and you can do the job is one of the best ways to actually get the job.

2. *You demonstrate that you already understand the job (even before you start) including not only the obvious (e.g., produce the architecture deliverables) but also its challenges.* Read and reread the job description and

try to find clues about hidden challenges. For example, perhaps the architecture team does not have a strong foothold in the organization or the senior management is not yet convinced the organization needs architecture (or another architect) because they do not really see the benefit. Some of these issues might come up in the interview and you have to steer your answers and demonstrate that you have successfully dealt with this kind of issue in the past and helped your organization to solve them.

Behavioral questions are a great way to show this. Many people look at them as a necessary evil but they are actually a great way to sell yourself by using the clues mentioned before to tailor your answers and demonstrate that you can deal with these kinds of situations. Nothing gets you closer to the job than planting the idea in your interviewers' heads that you can do the job from day one. One might ask how you can do this as you do not know anything about the team, the culture, and (not much) about the organization. However, none of the other candidates (unless they are internal, which is a different topic) know anything about these either, so even knowing a little bit or inferring this from these clues will put you in front as the main contender. Even if you do not have a lot of architecture experience, read a lot and try to understand how the problems you dealt with can be translated into architectural problems. For example, if you are a lead developer and you designed a certain application, stretch your thinking and be prepared to show how you can provide the architecture for the entire system based on the nonfunctional requirements.

3. *You demonstrate that you are a good fit for the team.* Again, some would argue that you cannot do this unless you know the team, but no matter who the team members are, there are certain things that would make you *desirable* to the team. The most obvious thing is that no one wants a grumpy teammate who is only interested in his or her success and does not collaborate well with others. Having conflicts is inevitable, but having them all the time is a sign that you might have a problem. One of the behavioral questions will most certainly be about a time when you had a conflict with someone (sometimes nicely hidden under the "had to work with someone difficult" form) and you have to provide an answer that shows that you can handle conflict in a professional manner. As with any behavioral question, the answer has to include a real example without focusing on the emotional side. My recommendation is the use of the STAR (situation, task action, result) format [3], describing the situation or task, the approach (key actions you took focusing on how you resolved the disagreement in a profes-

sional and productive way), and the results [happy ending: description of the positive outcome(s) of your actions]. This would make the interviewers understand that you have resolved quite a few conflicts in a diplomatic way and that therefore you are the best candidate to quickly integrate into the team.

One of the worst things is to give the impression that you are one (even if you are) of the geeks who is perfectly able to produce deliverables in his or her architecture ivory tower (isolated from the rest of the project team) but has real problems interacting with others in the team or organization. Somebody who cannot collaborate will only bring problems to the team and organization and will not be able to do the job properly. As explained previously, the architect role is not about being a hands-on coder who sits at his or her desk the whole day but involves collecting information from various stakeholders and selling your product (the architecture) back to them.

4. *You are able to adapt quickly to job specifics.* This might sound like a reiteration of a previous bullet, but, again, there might be some ways to put yourself at the front:

   • The organization might use a home-grown architecture framework or methodology. During the behavioral part of the interview (or whenever you might have a chance), you have to show that you can quickly adapt and be productive using that specific framework, eventually giving examples of how you already did this in the past.

   • The organization might use legacy applications and technologies and they are looking to you to help them to migrate to a new platform. Again, you should be able to emphasize that you can quickly adapt to this environment, assess the situation, and make a plan to bridge the gap. Give examples of how you did this in the past even if it might be about a senior developer job when you had to migrate from an old programming language to a new one or if you have an infrastructure background when you had to migrate from an old vendor product to a new one.

   • The organization might have a custom architecture process that is not efficient and needs to be improved. This might be about the way the architecture governance (including compliance, waivers, and so on) is enforced or about architecture checkpoints. For each of these cases, you must come up with examples of how you were able to improve similar processes in your previous jobs. This might be about the architecture or about improving the code review process or the business requirements collection.

5. *You can make a connection with your future boss.* You two actually click and he or she feels that you are quite familiar. This is another big advantage as you actually break the "new guy" barrier. Try to empathize with your future manager (he or she might be or have been an architect in the past and might have experienced similar problems) and to "walk for a day" in his or her shoes (do his/her work and experience his/her work-related problems). What is that you two have in common and what would bring you closer? It might be about trivial things like common hobbies, but usually it is about working in the same domain for years and being able to understand the joys of the job but also the pains you have to go through.

6. *You have the training/certifications and you keep yourself updated on the latest trends in architecture including new frameworks, methodologies, best practices, and patterns.* You may note that this is at the end and this placement is deliberate. As mentioned in many other places in this book, being an architect is not only about having the technical knowledge and demonstrating this by taking various certifications but also about having the soft skills required for you to succeed, such as leadership, communication, negotiation, and more (for a complete list, refer to Chapter 5). Obtaining a certification would most certainly put you in front, but the other factors mentioned before will usually have a bigger weight in the hiring decision. To sum up, my advice is to take certifications but not to see them as the most important factor. In my experience, I have seen people who were doing a lousy job as an architect but were continuing to pile up certification after certification thinking that the more they got, the better they would do as architects.

Excelling, or even doing a little bit better, at any of the factors enumerated before would most certainly help you to get the job. There are a lot of other factors that could influence the final decision (such as the hiring manager already having a preferred candidate), but you can only do your best to impress and to convince the interviewers of what a great asset you can be for the organization in the architect role.

10.2.2 The Real Cherry on Top (Do This and You Will Get a Job—Guaranteed)

Let us say that you studied and improved your knowledge, started to do a little bit of architecture (formally or informally), and read the guidelines in the previous section but you still cannot transition to an architect job. The last section of this chapter is trying to emphasize some of the critical factors that you can work on to improve your chances to become one:

- *Perseverance (again: do not lose confidence in yourself):* For any IT job positions but especially for architect roles, you will always find people trying to discourage you. You will always find people who will try to convince you that it is better to stay in your current job because becoming an architect is too difficult and does not offer a lot of professional rewards. This applies to the initial transition but also later on when you try to move to a different kind of architect position (like moving from an application to a systems architect). You must remember that no one is born an architect, but they understood what they needed to know to be an architect and got the knowledge by studying and the experience by applying that knowledge in real initiatives. They started small and evolved and you can do the same.

  You have to assess yourself with a critical eye, understand your level of knowledge and expertise, and then stick to this assessment no matter what others might say to you. The important word in the previous sentence is *critical.* There is no point in giving yourself a good mark when you know that you are not at that level. The idea is that your assessment should be as close as possible to your market value that others would establish (usually during an interview). You may stumble upon unfair assessments usually done by interviewers who do not have a clear idea of the responsibilities of the architect job (hands-on coding for an application architect, low-level design of an application for an enterprise architect, and so forth), and in these cases you have to stick to your self-assessment and to the definition of the role that you know based on the information in this book as well as other books, blogs, and articles.

  If you will forgive the personal anecdote, I remember years ago going to a consulting (senior) systems architect interview, going through human resources, interviewing with the direct manager and the subject matter experts (SMEs), and then failing at the interview with the vice president who was convinced that, at that level, I should be able to do code review and be very familiar with Java programming subtleties. Luckily for me, I was interviewing for another job in parallel and I ended up getting an offer in a few days. The bottom line is to never stop believing in yourself and your capability to be an architect. The moment you do, you actually stop the search to get an architect job.

- *Hone your skills, especially your soft skills.* Once more, the architect role is not only about having the technical or business knowledge required to do the job. It is also about having soft skills such as communication, leadership, and negotiation skills (for a complete list, see Chapter 5).

Without these, you will not be able to be a complete architect or even able to do the job properly.

When preparing yourself for this new career, you must work on technical and business knowledge, but also on how to become a better communicator (verbal, presentation, written), a better leader and negotiator (lead the team, work with the subject matter experts (SMEs), and sell them your final product: the architecture), and more. The self-assessment mentioned earlier should apply to soft skills too. If you know that you are a technical geek but you do not communicate well with people and are not good at influencing them to accept your proposed architecture, then you should not cover this up by saying that you are a technical guru. You must understand that no matter how good you are from a technical point of view, you need to become a good communicator, leader, and presenter to succeed as an architect. Do not forget: evaluate yourself with a critical eye and make a plan on how to get better. This might imply taking a training course, becoming a member in one of the Toastmasters clubs, or any other way that you can learn techniques and get some chances to practice and hone your skills.

- *Knowledge is good, but getting certified is better.* In a previous chapter, we discussed the importance of getting certified as an architect. The profession is still young and there are not many standard certifications (like the project management professional (PMP) one for a project manager) but you can still get a certification for a certain framework or methodology. There are quite a few certifications that belong to different vendors (Sun Certified Enterprise Architect, now Oracle, is one of the oldest that is focused on the Java Enterprise platform; others include EA certifications from Alcatel, BCS, Brocade, CheckPoint, Cisco, EMC, HP, IBM, Salesforce, SAP, and Red Hat), but most of them are platform or framework agnostic. These certifications are specific to each type of architect and the most common (and not vendor specific) credentials include:

  - *Open Certified Architect:* Competency-based certification (closest to the PMP one for the PM). This is applicable for enterprise, business, or IT architecture and requires the applicant to demonstrate real-world experience as opposed to the knowledge base alone (for more information, refer to http://www3.opengroup.org/certifications/professional/open-ca).

  - *TOGAF certification:* Widely recognized as the best EA certification to hold as the TOGAF framework/methodology has been adopted in 80% of the Fortune 500 companies. The certification is split in two parts: the first part (the foundation), which is more suitable for those entering the profession, and the full certification, which includes scenario-based questions that try to assess the experience of the applicant [4].

- *CITA:* IASA's Certified IT Architect [5] (http://iasaglobal.org/certifications/). A multilevel certification program known as the Certified IT Architect that is offered at foundation, associate, specialist, and professional levels. The CITA-S Specialization credential provides focus in one of four areas: business, infrastructure, information, and software architecture.

- *Zachman Certification:* Integrates learned theory with real-world experience. There are three levels of certification that go from Level 1 (defined as EA associate) to Level 2 (EA professional) and Level 3 (EA instructor) [6].

- *Certified SOA Architect:* Has a declared and definite emphasis on service-oriented architectures (SOAs), along with related technology solutions and infrastructure. It requires mastering the design of the service-oriented technology architecture, development, and delivery of working SOA solutions and the ability to integrate SOA solutions into general IT infrastructures. It is mostly geared towards software/ application architecture.

- *Archimate:* A modeling language certification adopted by TOGAF as the EA modeling language and supported by different vendor tools [7].

With so many options available, it is not easy to decide which one to pursue. Each one comes with a different background, experience, and aspirations, so one certification cannot be recommended over the others. This decision is not an easy one because it involves understanding the curriculum, learning, and taking the exam. Before you get into the certification process, you may have to evaluate which one is the best for you based on your experience and background, the type of architect job you would like to get, the fact that you would like to get the certification to help you in your current job (e.g., for a specific framework, and so on), or you want a certification that applies to most of the architect type positions out there on the market. Another factor that differentiates the certifications is if you are looking for a vendor-specific one (like the ones enumerated above) or an industry-recognized one (like Open Certified Architect or TOGAF). No matter what your final decision is, you have to start reading the material, register yourself on forums, and try to take advantage of the experience of others who already took the certification as well as taking as many practice tests as you can find on the Internet.

- *Know how to sell yourself.* No matter how good you are, how vast your experience is, or how many certifications you might hold, an important skill in making sure you get offered the job is knowing how to sell yourself. As with any other asset on the market, there is a supply-and-demand factor for each type of architect. Furthermore, there tends to be a higher demand and lower supply the further you move away from domain architectures (such as application, data, and infrastructure) and the closer you get to enterprise architecture. No matter what kind of architect job you seek, selling your skills definitely puts you in front of the crowd of applicants.

- *Do not be overconfident and sell skills or experience that you do not have.* Alternatively, do not forget that interviewers usually do not know you, so you must be prepared to tell them about your skills and experience. The best opportunity for this is during the behavioral part of the interview. This is the time when the interviewers are trying to get an understanding of how you have applied your skills to succeed in your job as an architect. This is an excellent opportunity for you to sell yourself not just by enumerating what kind of skills and experience you might have but also by giving examples of when (which previous job, initiative, or project) you used them to deliver successfully in the past.

Your interviewers are using an old principle that says that previous experience is a strong indicator of future performance. This is the moment to gently remind them about your certifications and about all your success in delivering. Have an example handy for each of the typical behavioral questions such as: "Tell me about a time when you had to communicate technically complicated material in such a way that it was clearly understood by a nontechnical person" or "Tell me about when you directed a discussion to build consensus and ensured that a common understanding was reached between all participants" (you can find lots of examples of behavioral questions; please refer to the bibliography for a list of sites). Many people have a lot of knowledge and experience, but they cannot sell themselves so they end up botching the interview and losing the job. This is also the employer's loss, but how can they know who you really are if you do not sell yourself?

As with any other job, you are the one who is responsible for getting the interview and becoming an architect. You will hear a lot of comments about people losing a job because they had bad luck and tough interviewers, because they did not have the experience, because the interviewers wanted a specific certification, or even because the interviewers had a preferred candidate. Some of these might be true, but no matter what the situation might be, if you do your homework and perform well, you have a much better chance to get it than when you are not clear on your objective and do not have a plan on how to achieve it.

References

[1] Architect Salary: United States, http://www.payscale.com/research/US/Job=Design_Architect/Salary, accessed May 2016.

[2] Glassdoor, https://www.glassdoor.ca/Salaries/, accessed May 2016.

[3] Pinola, M., "Use the STAR Technique to Ace Your Interviews," http://lifehacker.com/5960201/use-the-star-technique-to-ace-your-interviews, accessed May 2016.

[4] http://www.opengroup.org/togaf/, accessed May 2016.

[5] IASA Certifications, http://iasaglobal.org/certifications/, accessed May 2016.

[6] Zachman International, http://www.zachmaninternational.com, accessed May 2016.

[7] Archimate Certifications, http://www.opengroup.org/certifications/archimate, accessed May 2016.

11

The Future of IT Architecture

No one want to embark upon the arduous road of switching careers without knowing for sure that their newly chosen next career has a chance of surviving for at least a few decades, if not an entire lifetime. After all, starting or changing careers is a difficult and time-consuming initiative that involves understanding what the new career is all about, what its benefits are, and what background is required including the skills, competencies, knowledge, and certifications needed for success. This chapter focuses on the future of information technology (IT) architecture and the architect job as well as trends from recent years.

Just as computers are here to stay, so also are IT architecture and IT architect roles. Without them, companies would be unable to stay competitive, nor could they so effectively streamline their business and IT activities. For more information about the benefits of IT architecture, refer to Chapter 3.

IT architects are relatively new to the marketplace. As with other jobs, these roles will continue to evolve and require new skills and competencies or evolution of existing ones. Later, this chapter will discuss in detail what new or modified roles are likely to appear on the market and what new or modified skills they might require. Architect roles combine a multitude of skills and capabilities; the definition for each role is bound to evolve over time, and each individual role can be combined in a number of ways to form new ones. Because technology keeps evolving at an increased rate, the alignment between business and technology remains a driving factor toward company success. Architects will stay in demand to engineer increasingly complex systems, such as the increasingly virtualized and digitized infrastructure becoming so typical these days.

11.1 Constraints

IT architecture emerged because enterprises needed some way to make order out of IT chaos. No one wants to hire expensive consultants or full-time architects if they bring no value or do not save money for the organization. As a result, the most important constraints driving the evolution of the IT architect, and for IT architecture in general, is to:

- Continue to provide value to the organization by increasing efficiency by consolidating and reducing replication, diversifying business and IT services and processes, and simplifying operations.

- Be able to quickly respond to changes in business direction and industry landscape.

- Reduce the cost of operating the enterprise by increasing integration and alignment between business processes and the application components that realize them.

A lot of companies still lack clearly defined business capabilities and processes; sometimes they even lack operating models for their businesses. This makes it hard to understand their connections to the IT world and how improving this foundation (application and infrastructure architecture domains) can steer the business in the right direction.

The evolution of the IT architect role is constrained by the evolution of the organization, IT, and business. It is also constrained by the standardization of responsibilities for each of the roles so that architect jobs can be commoditized across various companies instead of being specific to the business and IT department of one company. At the enterprise architecture level, one conventional constraint is to think about architecture as having totally separate IT and business departments, each with its own strategies, objectives, and ways to measure how they are achieved. This occurs mostly for historical reasons (a few decades ago, there was no IT department) and is perpetuated by separating business and IT all the way up to the C-suite (corporate's most important senior executives) level. As the trend to eliminate boundaries between business and IT increases, the role of the architect will become more important in linking them together. The enterprise architect will become an important stakeholder who can take business strategies, goals, and current state as inputs and create roadmaps that present the transformation of the organization from the current to the target state clearly.

11.2 Drivers for Change

By looking at the most important drivers that will have an impact on architecture talent needs for the organization, we will be able to glimpse how architect roles will change. We can also speculate on what new architect roles will be created and what new responsibilities these new roles will require. The most significant organizational drivers to shape architecture should be:

- *Information over process:* Instead of focusing on business process automation and design, the competitive advantage provided by IT will need to focus on the customer experience, data analytics and knowledge worker enablement. This will require combining information, technology, and business architecture to enable business capabilities. For example, big data technology was born out of the necessity to analyze data to improve the way that businesses operate. Banks and credit card companies created enterprise data hubs for collecting data to analyze and determine patterns in client behavior. Understanding these patterns allowed them to adjust accordingly to increase revenue and improve their clients' experiences. These companies realized that they were sitting on a goldmine (the data that they collected over the years) and used technology to enable rapid and advanced analytics.

- *IT embedded in business services:* Applications and infrastructure will be centralized and embedded in the business services delivered by conjoined business and IT departments. Business services will be managed end-to-end from service strategy, design, transition, and continuous improvement. This will not eliminate the need for the application architects to design application services that can be reused across different lines of business, but it will increase the need for solution or enterprise architects to provide end to end solutions starting from the business and going down to infrastructure if needed. Solution architecture documents will include a business context view where business capabilities are aligned with corresponding application services.

- *Externalized delivery of services:* Commoditized services will come from external vendors. The organization will continue to focus on core services that provide competitive advantage while enabling services will come from vendors or providers. Internal resources will integrate services instead of developing and owning them. With the advent of cloud architecture, for example, companies can shift their focus and expense from buying infrastructure to support business applications to developing applications starting from a platform as a service (PaaS) or infrastructure as a service (IaaS) provided by various vendors. Financial companies in the United States and Canada have already started to use private clouds with IBM's Bluemix as the underlying PaaS. This enables rapid development of IBM Cloud Orchestrator as an IaaS for provision of virtual servers. As a result, architect jobs might also migrate to cloud providers. At present, there are great opportunities for cloud or big data architect roles.

- *Partnership with the business:* Business unit leaders will play a vital role in delivering services and managing technology. Increased emphasis on collaboration and information sharing will help speed time to market. Such collaboration will involve flexible virtual teams that can work seamlessly across geographical regions. The artificial separation between business and IT departments will slowly disappear as business dependency on evolution and innovation in the technology domain continues to increase. There is an increasing trend to run projects using Agile methodology. For this type of project, businesspeople are joining developers, analysts, and architects to iteratively gather needs, design architecture, and then implement it. As a result, the application/solution architect role will evolve to not only develop the target state but also build incremental/transition architectures to help the business understand changes on the technical side. Such architects will also help the team by breaking the big vision into smaller, more digestible, and understandable packages of changes.

- *Stand-alone and "ivory tower" roles will disappear:* IT roles, including architects, will be embedded in common business shared services groups instead of filling a separate IT function that encompasses strategy and governance. We will see less architecture without connection to business values and objectives even at the application or infrastructure levels. Architecture has to show its value at any and all of these levels. The first question an architect should ask is: "How can I help to move things forward?" Moving the company in the right direction and making sure that each project provides incremental change to reach the target state should be translated into daily steps ensuring that team members work to bring the project closer to its strategic goals. All this is impossible without collaboration. That is why the "ivory tower" will be gone and the architect must come down to Earth, rolling up his or her sleeves to help out wherever it will advance progress to meet business goals and objectives (for more information, please refer to http://www.gitshah.com/2011/01/ivory-tower-architecture.html)

- *Regulations:* In the United States, the use of enterprise architecture in the public sector has increased significantly owing to regulations mentioned in the Chapter 6. Federal, state, and local levels will increase their integration making it easier to provide better services for clients but also helping the fight against terrorism. For example, homeland defense and the effectiveness of the emergency response from various government organizations are largely dependent on information exchange between government levels.

  There should be an agreement on the architectural frameworks and methodologies that will have a positive effect on standardization of the architect roles as well.

These drivers for organizational change will create structural organization changes focused on closing the rift between the business and IT departments, connecting every architectural activity to business values and objectives.

11.3 New or Changed Architect Roles

The drivers listed above require not only minor changes in the responsibilities for various architect roles but also the creation of new roles. Both new and old roles might assume either brand-new responsibilities or they might take on some of the responsibilities that previously belonged to other roles.

Some of these new (or modified) roles will be relocated either to different departments within the enterprise or to a different enterprise altogether (such as cloud architects moving from financial companies to cloud providers). The tendency to include architecture teams in the business will increase as business leaders take a bigger role in managing the relationship with technology as well as technology itself. Some roles will migrate to new departments within the enterprise that sit between business and IT, focusing on the business and IT alignment and how well technology (applications and infrastructure) helps them to achieve business goals and strategies. The roles that might migrate to this new department include the enterprise architect and service architect. Following the outsourcing trend we have seen in recent years, certain low-level architecture roles might be externalized to software or infrastructure vendors (such as the software/application or infrastructure architect roles).

The traditional architect roles we discussed so far in the book might change as follows:

- *Enterprise architect:* This role will see an increased focus on the alignment of the business capabilities, services, processes, and enterprise applications and infrastructure. Currently housed in IT departments, these professionals might migrate to multifunctional departments that provide a mix of business and technology experts. Then business knowledge will be even more important because enterprise architects will need to understand the high-level business picture and also how technology can become an enabler of business strategies. This role will also need to understand and be able to design the people aspect of the organization, including the skills and responsibilities required for a certain business capability and how these evolve over time. The role will provide the primary liaison with the business helping the business leaders to understand and manage the risk exposure either due to the design, implementation, deployment, or maintenance issues of the applications automating the business processes or due to the outdated infrastructure relying on vendor products at the end of life.

In the financial industry, for example, this role is more of an enterprise solution architect, focusing on the end to end solution, taking into consideration established enterprise roadmaps, strategies and standards. An end-to-end focus means that the enterprise architect needs to focus not only on the evolution of the business capabilities through the use of technology but also on the personnel component. As an example, if the business strategy is to decrease time to market for products, this can be translated to the use of cloud architectures which speed completion through rapid provisioning of virtual servers and a platform ready environment. The enterprise architect has to plan for the ramp-up time required for the various roles (application/solution architect, developer, and analyst) but also for establishing strategic partnerships with cloud providers (such as with IBM to use Bluemix or Cloud Orchestrator) to provide knowledge and training.

- *Business architect:* This role will grow to not only develop business architecture for a certain business unit but to help to develop business strategies, goals, and objectives. The importance of this role will increase and the people who perform it will gain the most job security. That is because they must combine industry-related skills (such as business process modeling and business analysis) with business knowledge specific to the enterprise. They should be able to provide a high-level business view of the enterprise but also drill down to multiple levels of business capabilities and business processes.

- *Information/data architect:* This role will grow with the broadening scope of the architecture to include a picture of all information taken as input, manipulated and destroyed in the enterprise. Although this role was initially developed to provide the information subject matter expert for the enterprise, it will evolve to provide input, contribute and interpret enterprise architecture models, and understand and provide guidance in designing and managing the business objects that will later be implemented as data objects at the application level. Lots of blogs and other material emphasize the importance of data as the core of the enterprise. New trends and technologies like cloud, big data, and Data as a Service (DaaS) will continue to influence the evolution of this role from a technology skills point of view. The role will not move to one of the new departments created in the organization but lower-level skills like database management and programming might be externalized to vendors.

- *Solution/systems architect:* This role will continue to evolve due to the adoption of cloud computing, which will require the migration of applications and infrastructure to the cloud. This also comes with an increased focus on integration and providing functionality as a service at the application and infrastructure level. The role will include new responsibilities such as change management and IT continuity management. The solution/systems architect role will stay in the IT department and will focus on understanding how technology solution satisfies business requirements. This role provides the subject matter expertise in regards to IT architecture. It may be required to provide support to the enterprise/business architect.

- *Software/application/infrastructure architect:* These roles are tightly coupled to technology evolution. Cloud computing, software as a service (SaaS) (software is licensed on a subscription basis and is centrally hosted), and IaaS (hosting virtualized computing resources over the Internet) lead to more companies outsourcing some of their systems (usually the ones that serve no critical business capabilities and do not automate systems that confer competitive advantages). This inevitably leads in the long run to some of the application/infrastructure architect jobs moving to the vendors providing such services. Not all jobs like this will disappear as the company will not outsource systems that provide competitive advantage. Usually architects for such systems grow with the company, because they are often the ones who designed those systems. That is why the company will keep them around: they provide *insurance* that the system will keep running smoothly and without any incidents. Working for a vendor involves working with multiple clients with diverse problems. For such professionals, their experience will increase faster. This means that consulting and soft skills must be honed to respond to this change. This is discussed further later on.

The drivers mentioned in the previous section of the book might also lead to the creation of new architect roles to respond to organizational changes or to align with new roles. Aligning the organization to the information technology infrastructure library (ITIL) service model (organizing the enterprise based on the services that it provides and describing how a service provider creates value for an internal or external customer) means that the role of service manager evolves to become responsible for delivering the service as a whole, including both applications and infrastructure. For nonessential systems, service managers will focus on integration of internal services and outsourced ones (cloud, platform, infrastructure, and services). As a result, the service architect role might be introduced as a "joined-at-the-hip" (working very closely together) companion for the service manager to integrate business, information, and technology to deliver a complete service.

Basically, this new role will be responsible for defining the service architecture including the business processes, information, technology, and people. That work will be based on business strategies and goals previously defined by the business/enterprise architect in collaboration with business leaders. This role might require totally new skills or new versions of skills already mentioned in previous chapters. The delivery of a service that provides a competitive advantage to the company would require the ability to analyze the market and the competitors to translate the impact of market changes into IT needs. Communication and collaboration soft skills must also evolve to allow effective communication and collaboration (and to provide leadership) to other people located in different geographical areas. The role is similar to an enterprise architect because it needs to consider all domain architectures but has more depth because it deals with business needs for one service instead of multiple services. Figure 11.1 is similar to the earlier T shape (Figure 4.1) that focused on the breadth and depth of an architect versus a developer or an infrastructure specialist.

The same concept applies here because the service architect needs more depth for his or her service. However, an enterprise architect still needs more breadth to provide the architecture of services provided at the enterprise level.

The changes that we will see coming for the architect roles depend a lot on the changes through which enterprises will go. However, those changes also depend on the increase in maturity of the practice itself: the maturity of architecture processes and interfaces with other enterprise practices like portfolio or change management. This translates to more skills required for an architect to be successful in his or her job.

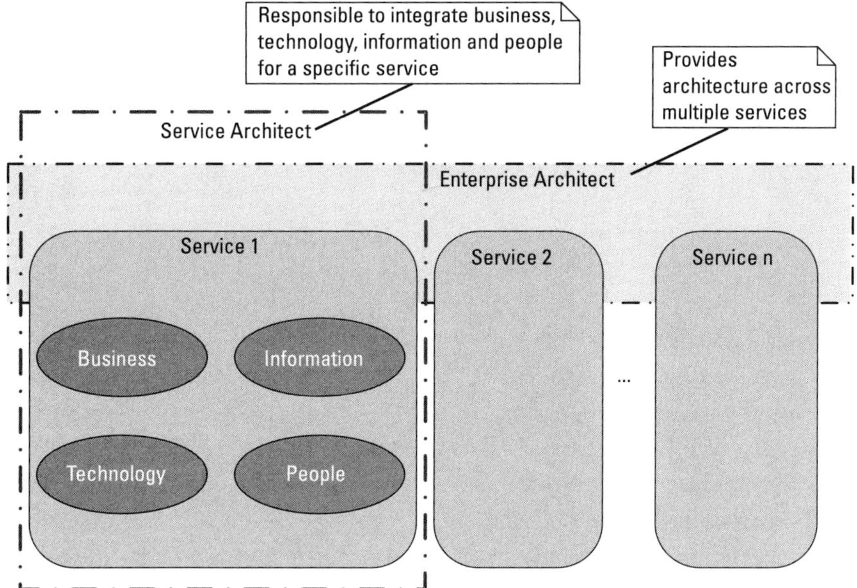

Figure 11.1 Service architect versus enterprise architect roles.

11.4 Skills

The constraints and drivers for change mentioned in the previous sections of this chapter will also influence the skills and responsibilities of the IT architect. For each of the architect job types, there are core skills that will remain essential to perform the job. These include architecture development, modeling (at various levels starting from the business and going down to information, application, and infrastructure), change management, development of business strategies, and scenarios.

The evolution of IT architect roles will also require new skills or the evolution of current skills to keep up with changes imposed by organizational changes. Some of the skills that will grow or evolve include:

- *Communication, influencing, and negotiation:* These soft skills usually go hand in hand (they are very closely related). Communication already started to evolve with the introduction of conferencing a few years ago. The concept of remote collaboration is central to outsourcing and has completely changed the way that the staff interacts to ensure delivery of the final product. What does this mean for an architect? It changes the way that the architect interacts with other practitioners to come up with a solution design (by involving subject matter experts to get the information required to put together the solution) but also to "evangelize" (build critical mass of support for the solution) and champion a solution through its implementation. The architect is must be able to communicate and influence stakeholders via phone or (best case) video conference. Reading body language and being able to adjust a presentation or the course of a meeting become more difficult when other participants are unseen and the only guide is the inflexions of voices. An architect also must negotiate the solution, navigate the treacherous seas of design shaping, and avoid or settle conflicts that often arise because of historical disagreements or misalignments between teams.

- *Strategy and governance:* These skills include the generation of the organization's overall objectives, principles, and strategies related to business and technologies in use in the organization. It was and will become an even more important skill for business and enterprise architects who should support the business and IT by defining the intentions of the organization at the business and technology level based on the goals and objectives usually set by the executives. The evolution of the governance skill will embrace more governance dimensions and domains (technology, IT, corporate, and so forth) and the ability to identify the interfaces between them.

- *IT financial management:* The ability to deal with the financial aspect of IT applications and infrastructure has become an important skill many types of architect. Starting from the business and enterprise architect who has to align the finance domain with business needs and continuing with the solution/service architect for whom the proposed solution should always take into consideration the financial aspect, this skill becomes a core differentiator in between successful and unsuccessful architects. For an enterprise architect, this is especially true as these architects must consider the enterprise as a whole. This translates into a few activities that maximize the value of the architect for the organization such as: analyzing the IT budget to help business and IT leaders to make better investment decisions as well as determine what initiatives will bring the most business value and provide competitive advantage.

- *IT service pricing:* The new service architect role will require estimation of the cost of IT services through various phases including service strategy, design, transition, and continuous service improvement phases. Taking the whole life cycle of the service into consideration, including not only the cost to deliver but also to continue to improve and adjust the service to better serve the needs of the business, will become a key skill in developing a complete service architecture.

- *Industry and market analysis:* This skill may sound more like one required for marketing or product development staff, but it will become indispensable for many types of architects (including enterprise, business, service, and solution). Gartner introduced the concept of *business disruptions* [1], which are unexpected and uncertain conditions that should be scouted for and adapted to. Their best example was Kodak, which filed for bankruptcy, although one of their engineers invented the digital camera long before this was recognized as a replacement for classic film cameras. Dealing with disruption continues to grow in importance as a skill. It has become essential in creating target states and roadmaps that take disruptions into account. The same discussion applies not only to the business but also to technology disruptions. The solution, application, or infrastructure architect should look for upcoming technology disruptions and shape the technical solution taking into consideration new trends in technology. They are the ones who should investigate these disruptions, making sure that they understand their importance, level of maturity, and impact on the organization and recommending changes to be implemented. This might also include market analysis, trying to identify trends and patterns that emerge either at the business or technology level. The architect cannot be the one deciding to go ahead and do something but should help make the risks of not dealing with disruption visible by developing risk and impact analysis and working with internal subject matter experts and stakeholders to identify the future-state that would address the disruption. No matter what the results of the analysis might be, the architect must realistically assess the organization's capacity to adapt and survive. He or she must also champion the implementation of the process to conduct regular assessments to identify disruptions.

- *Integration:* Although not a new skill, it becomes more important at all architecture levels. Starting from the integration of the applications and the integration of information, architects must be able to integrate information from multiple sources and channels. They must also see to the integration of business information to create a coherent view of the enterprise across all lines of business. More non-IT (financial, telecom, and so on) organizations try to cut cost and remain competitive by integrating commercial-off-the-shelf (COTS) applications used for auxiliary services with developed-in-house (DIH) applications that provide IT services for core functionality. To do this, the organization needs better integrators who can thrive in this mixed (and often mixed-up) environment to get the most for the business within their allocated budgets.

Cloud technology has become a major technical disruption. Many organizations have moved to adopt it sometimes without creating a plan for the integration of existing applications with cloud-based ones. At the enterprise level, integration encompasses the standards, policies, methodologies, and technologies that allow diverse applications to share information and functionality.

As we saw in the previous section, some architect roles might move to external vendors that provide a service. As a result, some of those skills will also become less important to the organization. They will most definitely not disappear but will instead be provided by external vendors that specialize in application or infrastructure solutions. One example is the application architecture skill. It will remain important for core applications that provide competitive advantage but in the future will be provided by the company to which a less vital service has been outsourced.

As IaaS gains more ground with the introduction of outsourced infrastructure, infrastructure architecture skills will also move to infrastructure vendors. Once the architect role becomes more formalized, low-level design including application, database, and infrastructure will be performed by domain specialists, allowing architects more time to focus on high-level designs. As the IT industry becomes more mature, organizations understand more that the architect is an expensive resource better to be used for architecture and not for an IT resource. Although architects may be good at everything from architecture and requirements gathering to coding, code review, and testing, they must stick to architecture if they are to deliver the value that an organization needs them to provide. This applies not only to technical skills but also to management skills around changes, incidents, problems, configurations, and technical resources.

As we have seen in previous chapters, the architect role sometimes has mixed responsibilities (especially for small to medium organizations). These can include managing projects (you have a high-level view of the tasks, so why not also manage the project?), business/systems analyst (you know a lot about the business and you know how to collect the architecture requirements, so why not collect the business requirements too?), or managing resources (you know the people, so why not manage them as well?). The trend to include responsibilities outside of architecture will continue for small organizations. However, it will most certainly decrease for medium and large organizations that understand this segregation of responsibilities and have the money to pay for more than one resource to perform multiple roles.

No matter what disruptions trigger changes at the organizational, business, or technology levels, the architect role will continue to grow as organizations began to grasp the complexity at the business and IT levels. This is most strongly driven by the need for a role to provide coherence across the various architecture domains. Roles might change and skills might evolve or adapt to new realities, but architect roles will continue to exist. The need for skilled architects has increased steadily from the days when the first architecture frameworks were introduced. They grew further when acts like Clinger-Cohen and Sarbanes-Oxley were passed and enforced by U.S. government technology in response to a number of major corporate and accounting scandals. They will continue to increase as more and more organizations start to see the benefits of creating and supporting dedicated architecture teams. There is still a tough road ahead, formalizing the responsibilities of the IT architect roles or even standardizing the role across the IT industry.

11.5   Final Considerations

We have now reached the end of the book. I hope that I have achieved the goals stated at its beginning: to provide extensive guidance on how to become an architect, as well as demonstrating the benefits of having an architecture team to provide high-level design for an organization at the enterprise, business, or technology level. If you are thinking about becoming an architect and you have a developer, analyst, infrastructure specialist, or similar background, the end of this book is the start of your road to becoming an architect. The first thing is to brush off your technical, business, and soft skills. Then you will want to acquire the recommended certifications and prepare for the interview once you think that you are ready to obtain an architect job. It is not an easy road because there is a lot of new knowledge to acquire, and there are a lot of ways that you may need to adapt your personality to become a successful IT architect.

One thing I can tell you for sure: it is totally worth it. I am not saying this job is a good fit for everyone. The purpose of the book was to provide you with the tools to first discover yourself and make sure that you make the right career choice and then to embark upon the path of becoming an architect. This could save you from the costly mistakes of engaging in a lengthy adventure when you do not even know what it is or if you will like it. I wish I had a book like this when I started my architect career. Unfortunately, there was no such book available, so I had to learn everything the hard way, getting the knowledge and shaping my personality through hard-earned experience to become successful. I read almost all the architecture books available. I found out which certifications are most desirable and what skills I really needed to polish. I also learned how to prepare myself for a successful interview.

I was also on the hiring side and I saw a lot of confusion in the industry, along with a plethora of varying opinions with regard to various architect roles. I also saw a lot of uncertainty about what kind of skills and experience one must inquire and how to assess individuals who claim the knowledge and experience to fill a certain architect role. I saw organizations in which the architecture function was there just to provide reports and to ensure compliance with acts and regulations. The management did not understand the architects' roles, so they were merely tolerated as exotic animals. These people were and are valuable resources that can be used for much more than that. However, as with any other change, architecture change has to be actively supported by executives and management who understand the value that the architects represent and how they can be utilized to push the company forward. I hope that this book managed to explain the importance of architecture and give you a hint of the ways that IT architecture can put your organization in front of the competition.

Architecture did not appear out of the need to task yet another team member with dreaming about a vision and championing new and cool technologies. Properly understood and used, architecture saves money and provides a steering wheel for your business that actually works. Architecture can even save a company by allowing its principals to understand business and technology changes before they begin to struggle and later on file for bankruptcy. As mentioned in *Enterprise Architecture as Strategy: Creating a Foundation for Business Execution* [2], you can standardize a business process and define how that process will be executed no matter who executes it. You can also increase integration by connecting the efforts across all lines of business to share data whenever and however possible. A good decision on the best operating model for the company is not an easy one: it requires deep analysis of the market and the way that the company can satisfy the market's needs with new or improved products.

Although this decision is taken by management, it is difficult to do so successfully without employing the expertise of an architect to identify the need for the diversification of products and to leverage economies of scale by introducing new products to the market. Some might even argue that this is a purely management or executive failure because they are paid to see such trends and react accordingly. But we are human beings and even though some of us might be able to see the future and steer companies in the right direction, most would benefit from being advised by the right person. This is especially true for a person who has the tools, frameworks, knowledge, and experience to provide such advice. At the enterprise level, this person is the enterprise architect, but providing the best advice also applies to each of the other architecture domains. The application architect has the tools and knowledge to help the company to build reusable components (therefore saving money), the data architect knows how to organize and make available data in the best way to serve the business, and the infrastructure architect knows how to navigate the maze of network devices, cloud architectures, and infrastructure services to ensure that the company can rely on the best infrastructure support.

Whether you are trying to change your profession or to use architecture as a meaningful tool for your organization, I hope this book provided you with the information you were looking for and I gladly welcome your feedback or comments (please send them to becomeITArchitect@gmail.com).

References

[1]   Judah, S., "Five Disruptions That Could Render Your Information Strategy Obsolete, and What to Do About Them," https://www.gartner.com/doc/3137419/disruptions-render-information-strategy-obsolete, accessed May 2016.

[2]   Ross, J. W., P. Weill, and D. Robertson, *Enterprise Architecture as Strategy: Creating a Foundation for Business Execution*, Cambridge, MA: Harvard Business Review Press, 2006.

Glossary

A complete glossary of information technology (IT) terms is out of the scope of this book; refer to the Gartner comprehensive glossary(http://www.gartner.com/it-glossary/) for IT terms or to the enterprise architecture glossary of terms (The EA Pad, Glossary of Terms, https://eapad.dk/ea3-cube/book/glossary-and-abbreviation-list/). Here is a summary of the most commonly used architectural terms for your reference. All Web sites were accessed in May 2016.

ADKAR   ADKAR is an acronym that represents the five milestones that an individual must achieve for change to be successful: awareness, desire, knowledge, ability, and reinforcement (Prosci, ADKAR model, https://www.prosci.com/adkar/adkar-model).

Application architecture   The layout of an application's deployment. This generally includes partitioned application logic and deployment to application server engines (Gartner Glossary, http://www.gartner.com/it-glossary/).

Archimate   An open and independent graphical modeling language used to represent an enterprise architecture. Archimate supports description, analysis, and visualization of architecture within and across business domains and is a technical standard from The Open Group based on ISO/IEC/IEEE 42010:2011.

Architectural description   A collection of artifacts used to describe and model an architecture and how the stakeholder's concerns are satisfied by that model (TOGAF 9.1, Architectural Artifacts, http://pubs.opengroup.org/architecture/togaf9-doc/arch/chap35.html).

Architectural model   This is a repository of enterprise, business, and technology entities, created using available standards, in which the primary concern is to illustrate a specific set of trade-offs inherent in the structure and design of a system ("The EA Pad: An Introduction to Enterprise Architecture," https://eapad.dk/ea3-cube/book/section1/chapter-1-an-overview-of-enterprise-architecture/).

Architecture building block (ABB)   This is a potentially reusable component of business, IT, or architectural capability that can be combined with other building blocks to deliver architectures and solutions (TOGAF 9.1, Building Blocks, http://pubs.opengroup.org/architecture/togaf9-doc/arch/chap37.html).

Architecture framework   Establishes a common practice for using, creating, interpreting, and analyzing architecture descriptions within a particular domain.

Architecture governance   The practice and orientation by which enterprise architectures and other architectures are managed and controlled at an enterprise-wide level (TOGAF 9.1, Architecture Governance, http://pubs.opengroup.org/architecture/togaf8-doc/arch/chap26.html).

Architecture principle   A qualitative statement of intent that should be met by the architecture. It has at least a supporting rationale and a measure of importance (TOGAF 9.1, Architecture Principles, http://pubs.opengroup.org/architecture/togaf8-doc/arch/chap29.html).

Architecture requirements   These set the boundaries and framework under which an architecture must operate and define basic IT requirements needed to support its business strategies, drivers and goals (TOGAF 9.1, ADM Architecture Requirements Management, http://pubs.opengroup.org/architecture/togaf9-doc/arch/chap17.html).

Artifact   An architectural work product that describes some aspect of an architecture (TOGAF 9.1, Architectural Artifacts, http://pubs.opengroup.org/architecture/togaf9-doc/arch/chap35.html).

Baseline (current view)   The current or as-is state of the business, information, or technology environment, captured in a set of graphic and textual models (TOGAF 9.1, Definitions, http://pubs.opengroup.org/architecture/togaf9-doc/arch/chap03.html).

Business architecture A description of the structure and interaction between the business strategy, organization, functions, business processes, and information needs (Enterprise Architecture Glossary by Set, http://www.cio. ca.gov/Government/IT_Policy/pdf/SIMM_58C_Enterprise_Architecture_ Glossary_04132011.pdf).

Business capability modeling A technique for the representation of an organization's business anchor model, independent of the organization's structure, processes, people, or domains (Gartner Glossary, Business Capability Modeling, http://www.gartner.com/it-glossary/business-capability-modeling/).

Business process management (BPM) The discipline of managing processes (rather than tasks) as the means for improving business performance outcomes and operational agility (Business process management, https://en.wikipedia. org/wiki/Business_process_management).

Business Process Modeling Notation (BPMN) A graphical language that serves as a basis for business processes and services modeling.

Commercial-off-the-shelf (COTS) software This is a product that is sold, leased, or licensed to the general public, offered by a vendor trying to profit from it, supported and evolved by the vendor who retains intellectual property rights and generally used without source code modification (Kazman, R., and D. Port, "COTS-Based Software Systems," Third International Conference (ICCBSS 2004) Proceedings (Lecture Notes in Computer Science), Springer, Redondo Beach, CA, February 1–4, 2004).

Concern A key interest that is crucially important to the stakeholders, addressed by an architecture view, and identified by an architecture description (TOGAF 9.1, Architectural Artifacts, http://pubs.opengroup.org/architecture/ togaf9-doc/arch/chap35.html).

Data architecture A description of the structure and interaction of the enterprise's major types and sources of data, logical data assets, physical data assets, and data management resources (TOGAF 9.1, Definitions, http://pubs. opengroup.org/architecture/togaf9-doc/arch/chap03.html).

Deliverable An architectural work product that is contractually specified and in turn formally reviewed, agreed, and signed off by the stakeholders (TOGAF 9.1, Definitions, http://pubs.opengroup.org/architecture/togaf9-doc/arch/chap03.html).

Engineering (nonfunctional) requirements Requirements that describe how the system should behave in contrast with functional requirements that describe what the system should do. Nonfunctional requirements impose constraints upon a system's behavior.

Enterprise The highest level (typically) of description for an organization that typically covers all missions and functions. An enterprise will often span multiple organizations (TOGAF 9.1, Definitions, http://pubs.opengroup.org/architecture/togaf9-doc/m/chap03.html).

Enterprise architecture (EA) A discipline for proactively and holistically leading enterprise responses to disruptive forces by identifying and analyzing the execution of change toward desired business vision and outcomes (Gartner IT Glossary, Enterprise Architecture (EA), http://www.gartner.com/it-glossary/enterprise-architecture-ea/).

Framework A style guide that defines the look, feel, and interoperability of software applications (Gartner IT Glossary, Framework, http://www.gartner.com/it-glossary/framework/).

Executive sponsor An executive who has decision-making authority over the EA program and who provides resources and senior leadership for the program ("The EA Pad: An introduction to Enterprise Architecture," https://eapad.dk/ea3-cube/book/section1/chapter-1-an-overview-of-enterprise-architecture/).

Gap The difference between a baseline/current state environment and a target state environment.

Governance The group of policies, decision-making procedures, and management processes that work together to enable effective planning and oversight of activities and resources.

Infrastructure as a Service (IaaS) One of the three fundamental service models of cloud computing alongside Platform as a Service (PaaS) and Software as a Service (SaaS). As with all cloud computing services, it provides access to computing resource in a virtualized environment, the cloud, across a public connection, usually the Internet. In the case of IaaS, the computing resource provided is specifically that of virtualized hardware, in other words, computing infrastructure (Interoute, "What Is IaaS?" http://www.interoute. com/what-iaas).

IEEE 1471 and ISO/IEC/IEEE 42010:2011   IEEE standards for describing the architecture of a software-intensive system and of the systems and software engineering.

Line of business (LOB)   A distinct area of activity within the enterprise. It may involve the manufacture of certain products, the provision of services, or internal administrative functions.

Methodology   A defined, repeatable series of steps intended to address some particular type of problem and typically centers on a defined process, but may also include certain definitions for content or deliverables (TOGAF 9.1, Definitions, http://pubs.opengroup.org/architecture/togaf9-doc/arch/chap03. html).

Pattern   A technique for putting building blocks into context. Building blocks are what you use, patterns can tell you how you use them, and when, why, and what trade-offs you have to make in doing so (TOGAF 9.1, Architecture Patterns, http://pubs.opengroup.org/architecture/togaf8-doc/arch/chap28.html).

Policy   The governing principle, plans, or rules that guide organizational behavior.

Reference model   An abstract description of a set of entities where those entities are described in terms of services and connectivity among such entities is loosely defined.

Roadmap   An abstract plan for business or technology change, typically operating across multiple architecture domains and over multiple years (TOGAF 9.1, Definitions, http://pubs.opengroup.org/architecture/togaf9-doc/arch/chap03.html).

Repository   An information system used to store and access architectural information, relationships among the architectural elements, and work products (Kappelman, L., The SIM Guide to Enterprise Architecture, Boca Raton, FL: CRC Press, 2009).

Service-oriented architecture (SOA)   TA design paradigm and discipline that helps IT meet business demands, producing interoperable, modular systems that are easy to use and maintain (Gartner IT Glossary, Service-Oriented Architecture (SOA), http://www.gartner.com/it-glossary/service-oriented-architecture-soa/).

Software as a Service (SaaS)   A software distribution model in which applications are hosted by a vendor or service provider and made available to customers over a network, typically the Internet (SaaS), http://searchcloudcomputing.techtarget.com/definition/Software-as-a-Service).

Solution architecture   A description of a discrete and focused business operation or activity and how information services/information technology (IS/IT) supports that operation, that typically applies to a single project or project release (Solution Architecture, https://en.wikipedia.org/wiki/Solution_architecture).

Stakeholder   A party that has an interest in an enterprise or project. The primary stakeholders in a typical corporation are its investors, employees, customers, and suppliers (TOGAF 9.1, Stakeholder Management, http://pubs.opengroup.org/architecture/togaf9-doc/arch/chap24.html).

Subject matter expert (SME)   A person who is an authority in some particular area or topic.

System   A collection of components organized to accomplish a specific function or set of functions.

Target architecture/target state   The description of a future state of the architecture being developed for an organization. A roadmap is usually developed to show the evolution of the architecture to reach or achieve a target state (TOGAF 9.1, Definitions, http://pubs.opengroup.org/architecture/togaf9-doc/arch/chap03.html).

TOGAF (The Open Group Architecture Framework)   The framework for enterprise architecture that provides an approach for designing, planning, implementing, and governing an enterprise information technology architecture.

Transition architecture   The formal description of one state of the architecture at an architecturally significant point in time (TOGAF 9.1, Definitions, http://pubs.opengroup.org/architecture/togaf9-doc/arch/chap03.html).

View   A representation of a whole system from the perspective of a related set of concerns (TOGAF 9.1, Developing Architecture Views, http://pubs. opengroup.org/architecture/togaf8-doc/arch/chap31.html).

Viewpoint   A specification of the conventions for constructing and using a view (TOGAF 9.1, Developing Architecture Views, http://pubs.opengroup. org/architecture/togaf8-doc/arch/chap31.html).

Zachman Framework   The Zachman Framework typically is depicted as a bounded 6 × 6 matrix with the communication interrogatives as columns and the reification transformations as rows. The framework classifications are represented by the cells, that is, the intersection between the interrogatives and the transformations (John A. Zachman and Zachman International, Inc., https://www.zachman.com/about-the-zachman-framework).

Bibliography

Alur, D., and J. Crupi, *Core J2EE Patterns: Best Practices and Design Strategies*, Upper Saddle River, NJ: Prentice Hall, 2001.

Alur, D., and D. Malks, *Core J2EE Patterns: Best Practices and Design Strategies*, 2nd ed., Upper Saddle River, NJ: Prentice Hall, 2003.

Banner, J., *Emotional Intelligence: Master the Art of Emotional Intelligence, Self-Awareness, and Relationship Skills (Communication Skills: How to Be a Leader, Boost Self Confidence and Win People Over)*, 2014.

Bauer, C., and G. King, *Java Persistence with Hibernate*, New York: Manning Publications, 2006.

Bell, J., *Emotional Intelligence: A Practical Guide to Mastering Emotions: Emotions Handbook and Journal (Emotions and Feelings)*, Amazon Digital Services LLC, 2014.

Bernard, S. A., *An Introduction to Enterprise Architecture*, 3rd ed., Bloomington, IN: Authorhouse, 2012

Block, P., *Flawless Consulting: A Guide to Getting Your Expertise Used*, 3rd ed., San Francisco, CA: Pfeiffer, 2011

Blosch, M., "Taking on Business Disruption Economic, Technical, Social," *Gartner EA Summit 2013*, 2013.

Booch, G., J. Rumbaugh, and I. Jacobson, *The Unified Modeling Language User Guide*, Reading, MA: Addison-Wesley Professional, 2005.

Booch, G., J. Rumbaugh, and I. Jacobson, *The Unified Software Development Process*, Reading, MA: Addison-Wesley, 1999.

Burke, B., "To the Point Measuring the Value of EA and Its Impact on Business Outcomes," *Gartner EA Summit 2013*, 2013.

Burton, B., "Business Capability Modeling Driving Business and Technology Change," *Gartner EA Summit 2013*, 2013.

Buschmann, F., et al., *A System of Patterns: Pattern-Oriented Software Architecture*, New York: Wiley, 1996.

Business Process Model and Notation (BPMN), Version 2.0, http://www.omg.org/spec/BPMN/2.0/, accessed May 2016.

Cade, M., and S. Roberts, *Sun Certified Enterprise Architect for J2EE Technology Study Guide*, Boston, MA: Sun Microsystems Press, 2002.

Chapman, S. G., *The Five Keys to Mindful Communication: Using Deep Listening and Mindful Speech to Strengthen Relationships, Heal Conflicts, and Accomplish Your Goals*, Boston, MA: Shambhala, 2012.

Clinger-Cohen Act (1996), Information Technology Management Reform Act, August 2006, http://dcmo.defense.gov/Portals/47/Documents/Clinger_Cohen_Act.pdf.

CMMI Product Team, Capability Maturity Model Integration (CMMI), Version 1.1, Staged Representation, CMU/SEI-2002-TR-029, ESC-TR-2002–029, Software Engineering Institute, 2002, http://www.sei.cmu.edu/reports/02tr029.pdf, accessed May 2016.

Cogen, P., *Communication: Communication Skills — Improve Your Communication Skills, Build Trust and Become Successful Now (Communication Skills in Relationships for Leadership, Social Skills, Leadership)*, Amazon Digital Services LLC, 2014.

Covey, S. R., *The 7 Habits of Highly Effective People: Powerful Lessons in Personal Change*, New York: Simon & Schuster, 2013.

Covey, S. M. R., R. R. Merrill, and S. R. Covey, *The SPEED of Trust: The One Thing that Changes Everything*, New York: Free Press, 2006.

Erl, T., *Service-Oriented Architecture: Concepts, Technology, and Design*, Upper Saddle River, NJ: Prentice Hall, 2005.

Erl, T., R. Puttini, and Z. Mahmood, *Cloud Computing: Concepts, Technology & Architecture*, Upper Saddle River, NJ: Prentice Hall, 2013.

Fowler, M., *Patterns of Enterprise Application Architecture*, Reading, MA: Addison-Wesley Professional, 2002.

Franch, X., "COTS-Based Software Systems," *4th International Conference*, ICCBSS Springer Berline Heidelbergy New York: Springer, 2005.

Gamma, E., et al., *Design Patterns: Elements of Reusable Object-Oriented Software*, Reading, MA: Addison-Wesley Professional, 1994.

Greefhorst, D., and E. Proper, *Architecture Principles: The Cornerstones of Enterprise Architecture*, New York: Springer, 2011.

Gutierrez, F., *Introducing Spring Framework: A Primer*, Apress, 2014.

Hedge, J., *The Essential DISC Training Workbook: Companion to the DISC Profile Assessment*, Redding, CA: DISC-U.org, 2012.

Hiatt, J. M., *ADKAR: A Model for Change in Business, Government and Our Community*, Loveland, CO: Prosci Learning Center Publications, 2006.

Hiatt, J. M., *Employee's Survival Guide to Change: The Complete Guide to Surviving and Thriving During Organizational Change*, Loveland, CO: Prosci Research, 2004.

Hilliard, R., IEEE Computer Society, IEEE Std 1471-2000: IEEE Recommended Practice for Architecture Description of Software-Intensive Systems, New York: IEEE, November 2000.

Hunter, D., *The Art of Facilitation: The Essentials for Leading Great Meetings and Creating Group Synergy*, San Francisco, CA: RHNZ Adult ebooks, 2009

International Chamber of Commerce, "Define Business Principles and Policies," http://www.iccwbo.org/Products-and-Services/Trade-facilitation/9-steps-to-responsible-business-conduct/Nine-steps/Steps/Step-4-Define-business-principles-and-policies/, accessed May 2016.

ISACA, "Template: Implementation and Migration Plan," http://www.isaca.org/Groups/Professional-English/enterprise-architecture/GroupDocuments/TOGAF%209%20Template%20-%20Implementation%20and%20Migration%20Plan.doc, accessed May 2016.

Jacobson, E., "Eric Jacobson on Management and Leadership," http://ericjacobsononmanagement.blogspot.ca/2010/09/good-sample-business-principles.html, accessed May 2016.

Jackson, M., *Presentation Skills Masterclass: Want to Be a Better Business Presenter? (Business Presentations and Public Speaking)*, www.ricksmithbooks.com, 2014.

Kappelman, L. A., *The SIM Guide to Enterprise Architecture*, Boca Raton, FL: CRC Press, 2009.

Kepner, C. H., and B. B. Tregoe, *The New Rational Manager, Updated Edition*, Princeton, NJ, 1997.

Kepner-Tregoe Inc., *Executive Problem Analysis and Decision Making*, Princeton, NJ: Kepner-Tregoe, 1973.

Koberg, D., and J. Bagnall, *The Universal Traveler: A Soft-Systems Guide to Creativity, Problem-Solving, and the Process of Design*, W. Kaufmann, 1974.

Kruchten, P., *The Rational Unified Process: An Introduction*, 2nd ed., Reading, MA: Addison-Wesley, 2003.

Lankhorst, M., *Enterprise Architecture at Work: Modelling, Communication and Analysis (The Enterprise Engineering Series)*, 3rd ed., New York: Springer, 2012.

Lindland, O. I., G. Sindre, and A. Sølvberg, *Understanding Quality in Conceptual Modeling*, New York: IEEE Software/Pearson, 1994.

Lockheed Martin Federal Systems, Odetics Intelligent Transportation Systems Division, ITS Implementation Strategy Example, 1998, www.iteris.com/itsarch/documents/imp/imp.pdf.

Malhotra, D., and M. Bazerman, *Negotiation Genius: How to Overcome Obstacles and Achieve Brilliant Results at the Bargaining Table and Beyond*, New York: Bantam, 2008.

McCloud, A., *Communication Skills: Discover The Best Ways to Communicate, Be Charismatic, Use Body Language, Persuade & Be A Great Conversationalist (Communication Persuasion, Body Language, Social Skills)*, Pro Mastery Publishing, 2014.

Millett, S., *Professional ASP.NET Design Patterns*, Indianapolis, IN: Wrox, 2010.

National Institute of Standards and Technology Integration Definition for Function Modeling (IDEF0) Draft, Federal Information Processing Standards Publication FIPSPUB 183. U.S. Department of Commerce, Springfield, Virginia, http://www.idef.com/wp-content/uploads/2016/02/idef0.pdf, accessed May 2016.

.NET engineering team, .NET Framework Blog, https://blogs.msdn.microsoft.com/dotnet/, accessed May 2016.

Noergaard, T., *Embedded Systems Architecture: A Comprehensive Guide for Engineers and Programmers*, Waltham, MA: Newnes, 2012.

Nygard, M., Michael Nygard, http://thinkrelevance.com/blog/2011/11/15/documenting-architecture-decisions, accessed May 2016.

Object Management Group, "UML Profile for Enterprise Distributed Object Computing Specification," http://www.omg.org/spec/EDOC/1.0/, 2004.

Object Management Group, "Unified Modeling Language (UML) Specification: Infrastructure, Version 2.4.1," 2003, http://www.omg.org/spec/UML/2.4.1/.

Ofni Systems, "Functional Requirements," http://www.ofnisystems.com/services/validation/functional-requirements/, accessed May 2016.

Palli, P., and G. Krishna Behara, "Wipro, Enterprise Architecture: A Practitioner View," *The Open Group Blog*, 2014, http://blog.opengroup.org/2014/09/17/enterprise-architecture-a-practitioner-view/.

Perks, C., and T. Beveridge, *Guide to Enterprise IT Architecture*, New York: Springer, 2013.

Pipeline Open Data Standard (PODS), http://www.pods.org/, accessed May 2016.

Rajshekhar, A. P., *.Net Framework 4.5 Expert Programming Cookbook*, Birmingham, UK: Packt Publishing, 2013.

Reed, P., "Reference Architecture: The Best of Best Practices," 2002, http://www.ibm.com/developerworks/rational/library/2774.html.

Reynolds, C., *Introduction to Business Architecture*, Boston, MA: Cengage Learning, 2009

Ross, J. W., P. Weill, and D. Robertson, *Enterprise Architecture as Strategy: Creating a Foundation for Business Execution*, Cambridge, MA: Harvard Business Review Press, 2006.

Sarbanes-Oxley Act (2002), http://www.soxlaw.com/, accessed May 2016.

Shupe, C., and R. Behling, "Developing and Implementing a Strategy for Technology Deployment," www.arma.org/bookstore/files/Shupe_Behling1.pdf, accessed May 2016.

Schekkerman, J., *Enterprise Architecture Good Practices Guide: How to Manage the Enterprise Architecture Practice*, Trafford Publishing, 2008.

Sessions, R., ObjectWatch, Inc., "A Comparison of the Top Four Enterprise-Architecture Methodologies," https://msdn.microsoft.com/en-us/library/bb466232.aspx, accessed May 2016.

Spewak, S. H., J. A. Zachman, and S. C. Hill, *Enterprise Architecture Planning: Developing a Blueprint for Data, Applications, and Technology*, New York: Wiley-QED Publication, 1992.

Stutzer, T., *Stakeholder Management*, Diplomica GmbH, 2003.

ter Doest, H. et al., "Viewpoints Functionality and Examples ArchiMate," https://doc.telin.nl/dscgi/ds.py/Get/File-35434, accessed May 2016.

Tyree, J., and A. Akerman, Capital One Financial, "Architecture Decisions: Demystifying Architecture," https://www.utdallas.edu/~chung/SA/zz-Impreso-architecture_decisions-tyree-05.pdf, accessed May 2016.

Van Haren, *TOGAF Version 9.1*, Berkshire, UK: Van Haren Pub, 2011.

Wilkinson, M., *The Secrets of Facilitation: The SMART Guide to Getting Results with Groups*, San Francisco, CA: Jossey-Bass, 2012.

Zachman, J. A., Zachman International, "Zachman Framework," 2003, https://www.zachman.com/about-the-zachman-framework, accessed May 2016.

About the Author

Cristian Bojinca is an enterprise/solution architect specializing in setting up architecture practice by identifying specific IT architect roles and tailoring architecture frameworks for the needs of the organization. He has developed enterprise roadmaps, connecting business drivers and strategies with technical capabilities, enabling the executive decision-making process, and setting up companies for success. He has worked with major financial organizations, providing architectural guidance to move them to a greater architecture maturity level, designing mission-critical, high-volume systems and applications. He is MSCE-, SCEA-, TOGAF-, and ITIL-certified, a member of the Association of Enterprise Architects (AEA), and various architecture LinkedIn groups where he has posted various articles.

Index

Managing Complex Technical Projects: A Systems Engineering Approach, R. Ian Faulconbridge and Michael J. Ryan

Managing Engineers and Technical Employees: How to Attract, Motivate, and Retain Excellent People, Douglas M. Soat

Managing Successful High-Tech Product Introduction, Brian P. Senese

Managing Virtual Teams: Practical Techniques for High-Technology Project Managers, Martha Haywood

Mastering Technical Sales: The Sales Engineer's Handbook, Third Edition, John Care and Aron Bohlig

The New High-Tech Manager: Six Rules for Success in Changing Times, Kenneth Durham and Bruce Kennedy

The Parameter Space Investigation Method Toolkit, Roman Statnikov and Alexander Statnikov

Planning and Design for High-Tech Web-Based Training, David E. Stone and Constance L. Koskinen

A Practical Guide to Managing Information Security, Steve Purser

The Project Management Communications Toolkit, Second Edition, Carl Pritchard

Preparing and Delivering Effective Technical Presentations, Second Edition, David Adamy

Reengineering Yourself and Your Company: From Engineer to Manager to Leader, Howard Eisner

The Requirements Engineering Handbook, Ralph R. Young

Running the Successful Hi-Tech Project Office, Eduardo Miranda

Successful Marketing Strategy for High-Tech Firms, Second Edition, Eric Viardot

Successful Proposal Strategies for Small Businesses: Using Knowledge Management to Win Government, Private Sector, and International Contracts, Sixth Edition, Robert S. Frey

Systems Approach to Engineering Design, Peter H. Sydenham

Systems Engineering Principles and Practice, H. Robert Westerman

Systems Reliability and Failure Prevention, Herbert Hecht

Team Development for High-Tech Project Managers, James Williams

For further information on these and other Artech House titles, including previously considered out-of-print books now available through our In-Print-Forever® (IPF®) program, contact:

Artech House
685 Canton Street
Norwood, MA 02062
Phone: 781-769-9750
Fax: 781-769-6334
e-mail: artech@artechhouse.com

Artech House
16 Sussex Street
London SW1V 4RW UK
Phone: +44 (0)20 7596-8750
Fax: +44 (0)20 7630-0166
e-mail: artech-uk@artechhouse.com

Find us on the World Wide Web at: www.artechhouse.com